v3-13   ree

WITHDRAWN

Damaged, Obsolete, or Surplus
Jackson County Library Services

DATE DUE

OCT 4 '00	sept 9-8-2011	
OCT 16 '00		
NOV 16 '00		
MAR 29 '0		
APR 18 '01		
MAY 12 '01		Oct10-17-2011
AUG 21 '01		
JAN 25 '02		
DEC 20 2005		
MAR 15 '11		

8/00

THE
TRIPLE HELIX

THE
TRIPLE HELIX

Gene, Organism, and Environment

RICHARD LEWONTIN

HARVARD UNIVERSITY PRESS

CAMBRIDGE, MASSACHUSETTS

LONDON, ENGLAND · 2000

Originally published as Gene, organismo e ambiente. Copyright © 1998 by Guis. Laterza & Figli Spa, Roma-Bari. English language edition arranged through the mediation of Eulama Literary Agency.

Library of Congress Cataloging-in-Publication Data

Lewontin, Richard C., 1929–
    The triple helix : gene, organism, and environment / Richard Lewontin.
      p. cm.
    Portions previously published as: Gene, organismo e ambiente. Roma : Laterza, 1998.
    Includes bibliographical references and index.
    ISBN 0-674-00159-1
      1. Molecular biology—Philosophy.   2. Developmental biology—Philosophy.
3. Ecology—Philosophy. 4. Evolution (Biology) I. Lewontin, Richard C., 1929–.
Gene, organismo e ambiente.   II. Title.

QH506.L443 2000
572.8'01—dc21   99-053879

ACKNOWLEDGMENTS

The first three chapters of this book were originally presented as lectures in the Lezioni Italiani in Milan, and I am very grateful to Pino Donghi of the Fondazione Sigma Tau for having arranged those lectures and for having pressed for their original publication as *Gene, Organismo e Ambiente* by Editori Laterza. I added the last chapter at the urging of Michael Fisher of Harvard University Press, whose suggestions are greatly appreciated. I owe an immense debt to Richard Burian, whose careful and sophisticated review of the manuscript, especially of Chapter IV, resulted in a critical improvement from an earlier version. Finally, the text was greatly improved by Camille Smith's light but unerring editing. I wish she could edit everything I write.

CONTENTS

I GENE AND ORGANISM

It is not possible to do the work of science without using a language that is filled with metaphors. Virtually the entire body of modern science is an attempt to explain phenomena that cannot be experienced directly by human beings, by reference to forces and processes that we cannot perceive directly because they are too small, like molecules, or too vast, like the entire known universe, or the result of forces that our senses cannot detect, like electromagnetism, or the outcome of extremely complex interactions, like the coming into being of an individual organism from its conception as a fertilized egg. Such explanations, if they are to be not merely formal propositions, framed in an invented technical language, but are to appeal to the understanding of the world that we have gained through ordinary experience, must necessarily involve the use of metaphorical language. Physicists speak of "waves" and "particles" even though there is no medium in which those "waves" move and no solidity to those "particles." Biologists speak of genes as "blueprints" and DNA as "information." Indeed, the entire body of modern science rests on Descartes's metaphor of the world as a machine, which he introduced in Part V of the *Discourse on Method* as a way of understanding organisms but then generalized as a way of thinking about the entire universe. "I have hitherto described this earth and generally the whole visible world, as if it were merely a machine in which there was

3

nothing at all to consider except the shapes and motions of its parts" (*Principles of Philosophy,* IV).

While we cannot dispense with metaphors in thinking about nature, there is a great risk of confusing the metaphor with the thing of real interest. We cease to see the world *as if* it were *like* a machine and take it to *be* a machine. The result is that the properties we ascribe to our object of interest and the questions we ask about it reinforce the original metaphorical image and we miss the aspects of the system that do not fit the metaphorical approximation. As Alexander Rosenblueth and Norbert Weiner have written, "The price of metaphor is eternal vigilance."[1]

A central problem of biology, not only for biological scientists but for the general public, is the question of the origin of similarities and differences between individual organisms. Why are some short and others tall, some fat and others thin, some prolific setters of seed and some nearly sterile, some clever and others dull, some successful and others failures? Every individual organism begins life as a single cell, a seed or fertilized egg, that is neither tall nor short, neither clever nor dull. Through a series of cell divisions, differentiations, and movements of tissues, an entire organism is formed that has a front and a back, an inside and an outside, and a collection of organs that interact with each other in a complex way. Changes in size, shape, and function occur continually throughout life until the moment of death. As we grow older we grow taller at first and then shorter, our muscles become stronger and then weaker, our brains acquire more information and then seem to lose it. The technical term for this life history change is *development,* and the study of the process is called *developmental biology* (or, in cognitive and behavioral studies, *developmental psychology*).

But the term *development* is a metaphor that carries with it a prior commitment to the nature of the process. Development (*svillupo* in Italian, *desarrollo* in Spanish, *Entwicklung* in German) is literally an unfolding or unrolling of something that is already present and in some way preformed. It is the same word that we use for the process of realizing a photographic image. The image is already immanent in the exposed film, and the process of development simply makes this latent image apparent. This is precisely the view that developmental biology has of the development of an organism. Modern developmental biology is framed entirely in terms of genes and cell organelles, while environment plays only the role of a background factor. The genes in the fertilized egg are said to determine the final state of the organism, while the environment in which development takes place is simply a set of enabling conditions that allow the genes to express themselves, just as an exposed film will produce the image that is immanent in it when it is placed in a chemical developer at the appropriate temperature.

One of the most important issues in the premodern biology of the eighteenth century was the struggle between the preformationist and epigenetic theories of development. The preformationist view was that the adult organism was contained, already formed in miniature, in the sperm and that development was the growth and solidification of this miniature being. Textbooks of modern biology often show, as an example of the quaint notions of past eras, a seventeenth-century drawing of a tiny homunculus packed into a sperm cell (see Figure 1.1). The theory of epigenesis was that the organism was not yet formed in the fertilized egg, but that it arose as a consequence of profound changes in shape and form during the course of embryogenesis. It is usually said that the epigenetic view decisively defeated preformationism. After all, nothing could seem to us more foolish than a picture of the tiny man inside the sperm cell. Yet it is really preformationism that has triumphed, for there is no essential difference, but only one of mechanical details, between the view that the organism is already formed in the fertilized egg and the view that the complete blueprint of the organism and all the information necessary to specify it is contained there, a view that dominates modern studies of development.

⮞ The use of the concept of development for the changes through which an organism goes during its lifetime is not simply a case of available language influencing the content of ideas. When it was decided to make an ancient language, Hebrew, into a modern one with a technical vocabulary, the word chosen for the development of an organism, *Lehitpateach,* was the same as the word chosen for the development of a film, but in the reflexive form, so an organism literally "develops itself." Moreover, the word *evolution* has the same meaning of an unfolding, and for this reason Darwin did not use the word in the first edition of the *Origin.* Before Darwin the entire history of life on earth was seen as an orderly progression of immanent stages. While Darwin freed the theory of this element of predetermination, its intellectual history has left its trace in the word.

What is reflected in the use of these terms is the deep commitment to the view that organisms, both in their individual life histories and in their collective evolutionary history, are determined by internal forces, by an inner program of which the actual living beings are only outward manifestations. This com-

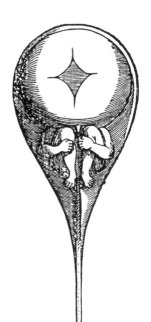

Figure 1.1. A picture by the seven-teenth-century microscopist Nicolaas Hartsoeker of the human sperm, show-ing it as containing a microscopic in-fant folded in a fetal position. This al-ready-formed infant supposedly grew larger during fetal development, with the mother's egg providing only the nu-trition for its growth.

mitment is an inheritance from the Platonic typological understanding of nature according to which actual material events, which may differ in varying degrees from each other, are the imperfect and accidental realizations of idealized types. The actual is the ideal seen "as through a glass, darkly." This was the view of species that was dominant until the twentieth century. Each species was represented by a "type" description, and an actual specimen was deposited in some collection as representative of the type, while all other individuals of the species, varying from the "type," were regarded as imperfect realizations of the underlying ideal. The problem of biology, then, was to give a correct anatomical and functional description of the "types" and to explain their origin. Modern evolutionary biology rejects these Platonic ideals and holds that the actual variation among organisms is the reality that needs to be explained. This change in orientation is a consequence of the rise of the Darwinian view that the actual variation among organisms is the material basis on which evolutionary change depends.

The contrast between the modern Platonic theory of development and Darwinian evolutionary theory is the contrast between two modes of explanation of the change of systems through time. Development is a *transformational* theory of change. In transformational theories the entire ensemble of objects changes because each individual object undergoes during its lifetime the same law-like history. The cosmos is evolving because all stars of the same initial mass go through the same sequence of thermonuclear and gravitational changes on their way to a predictable position in the main sequence. As a group, seventy-year-olds are grayer and more forgetful than thirty-five-year-olds because all the individuals have been aging in body and mind. In contrast, the Darwinian theory of organic evolution is based on a *variational* model of change. The ensemble of individuals changes, not because each individual is undergoing a parallel development during its life, but because there is variation among individuals and some variants leave more offspring than others. Thus the ensemble changes as a whole, by a change in the proportional representation of the different variants, which are themselves unchanging in their properties. If insects are becoming more resistant to insecticides, it is not because each individual is acquiring greater and greater resistance during its lifetime, but because the resistant variants live and reproduce while the susceptible organisms are killed.

A consequence of the difference between these two models of change is a difference in the problematic of biological disciplines that incorporate them. For evolutionists the differences between individual organisms and the differences between closely related species are at the center of attention. The variation is the primary object of enquiry. Its causes need to be explained and it needs to be incorporated into the explanatory narratives of the origin and evolution of species. Similarities between organisms are taken to be largely historical consequences of common ancestry, of the expected similarity between close relatives, rather than as consequences of functional laws. Indeed the entire science of systematics, whose purpose is to reconstruct the relationships and ancestry patterns of species, uses as its only data the observed patterns of similarity.

In contrast, for developmental biologists the variation between individual organisms, and even between species, is not of interest. On the contrary, such variation is an annoyance and is ignored wherever possible. What is at the center of interest is the set of mechanisms that are common to all individuals and preferably to all species. Developmental biology is not concerned with explaining the extraordinary variation in anatomy and behavior even between offspring of the same mother and father, which enables us to recognize individuals as different. Even the large differences between species are not within the concerns of the science. No developmental biologist asks why human beings and chimpanzees look so different, except to say the obvious: that they have different genes. The present agenda of developmental biology concerns how a fertilized egg becomes differentiated into an embryo with a head at one end and an anus at the other, why it has exactly two arms at the front and two legs at the back rather than six or eight appendages projecting from the middle of the body, and why the stomach is on the inside and the eyes on the outside.

The concentration on developmental processes that appear to be common to all organisms results in a concentration on those causal elements which are also common. But such common elements must be internal to the organism, part of its fixed essence, rather than coming from the accidental and variable forces of the external milieu. That fixed essence is seen as residing in the genes.

⮞ One of the most eminent molecular biologists, Sydney Brenner, speaking before a group of colleagues, claimed that if he had the complete sequence of DNA of an organism and a large enough computer then he could compute the organism. The symbolic irony of this remark is that it was made in his opening address of a meeting commemorating the one hundredth anniversary of Darwin's death.[2] A similar spirit motivates the claim by yet another major figure in molecular biology, Walter Gilbert, that when we have the complete sequence of the human genome "we will know what it is to be human."[3] Just as the metaphor of development implies a rigid internal predetermination of the organism by its genes, so the language used to describe the biochemistry of the genes themselves implies an internal self-sufficiency of DNA. First, DNA is described in textbooks and popularizations of science as "self-replicating," producing copies of itself for every cell and every offspring. Second, DNA is said to "make" all the proteins that constitute the enzymes and structural elements of the organism. The project to characterize the entire DNA sequence of humans has been called by molecular biologists "the search for the Grail," and the metaphor of the Holy Grail seems entirely apt since it too was said to be self-renewing (although only on Good Friday) and all-sustaining, providing nourishment for those who partook of it "sans serjant et sans seneschal," without servant or steward.

The metaphor of unfolding is then complete from the level of molecules to the level of the whole organism. Molecules that reproduce themselves and that have the power to make the substances of which the organism is composed contain all the information necessary to specify the complete organism. The development of an individual is explained in standard biology as an unfolding of a sequence of events already set by a genetic program. The general schema of developmental explanation is then to find all the genes that provide instructions for this program and to draw the network of signaling connections between them. The ultimate explanatory narrative of developmental biology will then be something like the following: "The division of the cell turns on gene A, which specifies a protein that binds to the DNA of the controlling regions of gene B and gene C, which results in an activation of these genes, whose protein products combine with each other to form a complex that turns off gene A in the cell near the surface but not in the cell that is more interior, which, etc., etc."

When this complete narrative finally becomes available, as it certainly will in the not too distant future for large parts of early embryonic development of worms and fruit flies, then the fundamental problem of development, as currently understood by the communal agreement of developmental biologists, will have been solved. Moreover, some of the elements of this narrative must be common not only to individuals who are examples of the same species ideal but to a vast array of species that are organized in similar ways. The greatest excitement in the study of development has been generated by the discovery that there are genes concerned in the ordering of the parts of an organism from one end to the other, the *homeobox* genes, that can be found in humans, insects, worms, and even plants. That such genes exist is undoubtedly of very great interest, especially to the evolutionist concerned with the underlying continuities in the history of life. For the program of developmental biology, however, the excitement arises from that discovery's embodiment of the ultimate program of the science.

A last feature of the unfolding model is that the life history pattern is seen as a regular sequence of stages through which the developing system passes, the successful completion of one stage being the signal and condition for passing on to the next stage. Differences in pattern between species and individuals are then thought of as the result of adding new stages or of "arrested development" in an earlier stage. The role of the external environment in this theory is twofold. First, some environmental trigger may be necessary to start the process. Desert plants produce seed that lies dormant in the dry soil until occasional rainfall breaks the dormancy and development of the embryo begins. Second, once the *déclenchement* has occurred, setting the process in motion, some minimal environmental conditions must exist to allow the unfolding of the internally programmed stages, just as the correct chemical baths are required for the development of a film but do not alter the shape of the final image.

The notions of regular stages as normal and arrested development as the source of the abnormal have been central to theories of psychological maturation, as in the Piagetian stages through which the child must pass to reach psychological maturity and the Freudian theory of fixation at infantile anal or oral erotic stages as a source of neurosis. Evolutionary explanation too has had its share of stage theories. The fetuses of humans and apes resemble each other much more than the adults do, and adult humans have morphological features that make them resemble fetal apes, for example in the shape of the skull and face. A generalization of these observations has led to the theory of *neoteny,* that there is a trend in evolution to be born earlier, cutting off development at an earlier stage in the ancestral developmental sequence.

But a contrary trend is also observed when even earlier embryonic stages are examined and a comparison is made with much more distantly related forms. The very young embryos of terrestrial vertebrates have gill slits like fish and amphibia, which then disappear in later development. This is an example of the rule that "ontogeny recapitulates phylogeny." Organisms that have appeared later in evolution seem to have added new stages to their development while still passing through the earlier ones of their ancestors, rather than losing them by neoteny.

At a previous time in the history of evolutionary theory, during the nineteenth century, these observed regularities were taken to be general causal properties of development and evolution, but they passed out of explanatory fashion during the rise of modern mechanistic biology because no mechanism could be found that would generate such regularities. With the discovery of homeobox genes they have been rejuvenated in a more sophisticated form. Figure 1.2 shows a change in *Drosophila* from an antenna to a leg as the result of a mutation in a homeobox gene. It had long ago been supposed, on the basis of comparisons of various arthropods, that antennae and legs were simply modifications of the same basic appendages. The mutation shown in Figure 1.2 is strong confirmation, at the genetic level, of this deduction. If all animals share the same deeply entrenched genetic program of anterior-posterior and dorsal-ventral differentiation, then it is easy to imagine how evolution may add and subtract stages of this common program by changes in gene signaling.

➢ The structure of explanation of development as an unfolding of a predetermined genetic program has powerful consequences for the explanation of the manifest variation among organisms. Although developmental biology is not primarily concerned with variation, the existence of variation among individuals enters into the program of investigation in a special way through the use of gene mutations that have drastic effects on development. The standard method for showing that a gene is important in, say, the development of wings in an insect, is to find a mutation of the gene that prevents wings from being formed or, even more interesting, that results in the formation

Figure 1.2. The effect of the mutation Antennapedia on the development of the head of *Drosophila.* Normal flies have an antenna consisting of small segments and a bristle-like extension. The mutant replaces this with a well-developed leg-like appendage, showing that the same basic developmental pathway leads either to an antenna or to a leg. Courtesy FlyBase/F. R. Turner. Used with permission.

of extra wings. The use of drastic gene mutations as the primary tool of investigation is a form of reinforcing practice that further convinces the biologist that any variation that is observed among organisms must be the result of genetic differences. This reinforcement then carries over into biological theory in general.

While observations of the natural variation between individuals are not taken into account in building the theory of development, the existence of such variation is obvious to all. Especially in the human species this variation may have great individual and social consequences. Differences in temperament, in the possession of particular physical and mental abilities, in health and disease, in social power all demand explanation. Up until the Second World War biologists, especially geneticists, were for the most part biological determinists who ascribed to genes the chief causal influence in molding social, psychological, and cognitive differences between individuals. Then, as the consequences of the biological theories of race and character in hands of the National Socialists became widely known, there was a general revulsion against biological determinism and it was replaced by a widespread environmentalist explanation of social facts. But this environmentalist dominance was short-lived, and within twenty years of the end of the war, genetic explanations again came to dominate, in no small part because psychology and sociology failed to produce a coherent predictive scheme for human psychic and social development.

The reigning mode of explanation at present is genetic. Reinforced by the observation that some human disorders result from mutation of clearly defined genes, nearly all human variation is now ascribed to genetic differences. From the undoubted fact that gene mutations like the Tay-Sachs mutation or chromosomal abnormalities like the extra chromosome causing Down syndrome are the sources of pathological variation, human geneticists have assumed that heart disease, diabetes, breast cancer, and bipolar syndrome must also be genetic variants. The search for genetic variation underlying widespread human disease conditions is a major preoccupation of medical research, a major consumer of publicly funded research projects, and a major source of news articles on health. Nor is it only pathological variation that is explained genetically. Variations in sexual preference, in school performance, in social position are also seen as consequences of genetic differences. If the development of an individual is the unfolding of a genetic program immanent in the fertilized egg, then variations in the outcome of development must be consequences of variations in that program.

The trouble with the general scheme of explanation contained in the metaphor of development is that it is bad biology. If we had the complete DNA sequence of an organism and unlimited computational power, we could not compute the organism, because the organism does not compute itself from its genes. Any computer that did as poor a job of computation as an organism does from its genetic "program" would be immediately thrown into the trash and its manufacturer would be sued by the purchaser. Of course it is true that lions look different from lambs and chimps from humans because they have different genes, and a satisfactory explanation for the differences between lions, lambs, chimps, and us need not involve other causal factors. But if we want to know why two lambs are different from one another, a description of their genetic differences is insufficient and for some of their characteristics may even be irrelevant. Even a very faulty computer will be satisfactory if one is only interested in calculations to an order of magnitude, but for accuracy to one decimal place a different machine is needed. There exists, and has existed for a long time, a large body of evidence that demonstrates that the ontogeny of an organism is the consequence of a unique interaction between the genes it carries, the temporal sequence of external environments through which it passes during its life, and random events of molecular interactions within individual cells. It is these interactions that must be incorporated into any proper account of how an organism is formed.

First, although internally fixed successive developmental stages are a common feature of development, they are not universal. A striking case is the life history pattern of certain tropical rain forest vines (see Figure 1.3).[4] After the seed germinates on the forest floor, the shoot grows along the ground toward any dark object, usually the trunk of a tree. At this stage the plant is positively geotropic and negatively phototropic. If it encounters a small log it grows over it, putting out leaves (form T_L), but then continues to grow along the ground without leaves (form T_S). When it reaches a tree trunk it switches to being negatively geotropic and positively phototropic and begins to climb the trunk away from the ground and toward the light (form A_A). As it climbs higher more light reaches its growing tip, and it begins to put out leaves of a particular shape at characteristic intervals along its growing stem. As it grows higher and yet more light falls on it the leaf shape and distance between leaves changes, and at a sufficient light intensity it begins to form flowers. If a growing tip grows out along a branch of the tree it becomes again positively geotropic and negatively phototropic, changes its leaf shape and spacing, and forms an aerial vine that grows down toward the ground (form A_D). When it reaches the ground it again returns to the T_S form until it encounters another tree, and there it may climb even higher in form A_A, as shown on the right in Figure 1.3. Each pattern of leaf shape, leaf spacing, phototropism, and geotropism is dependent on the incident light conditions, and there is no internally fixed order of stages. Even the description of the stages is somewhat arbitrary, since the shape and spacing of leaves change continuously as the stem ascends the tree trunk.

Figure 1.3. Changes in the morphology of the tropical vine *Syngonium* as it grows. T_L and T_S are terrestrial patterns, A_A is the pattern as it ascends a tree, A_D is the pattern when it is descending from a branch toward the ground.
From T. S. Ray, "Growth and heterophylly in an herbaceous tropical vine, Syngonium (Araceae)" (Ph.D. thesis, Harvard University, 1981).

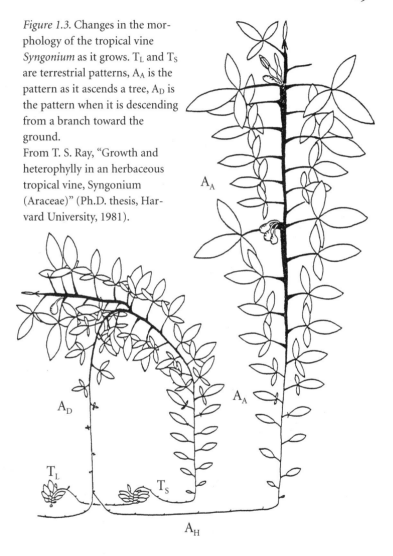

It might be that such switching among growth patterns under the influence of environment would be possible only in plants, because they have embryonic tissue at their growing points throughout their entire lives. However, the same phenomenon can be seen in the regulation of differentiation in insects. The wing of a moth develops from a lump of tissue, the wing imaginal disc, during the development of the adult inside the pupal case. The wing imaginal discs are generally considered to be independent of the discs that develop into the head or legs or abdomen or genitalia. Nevertheless, if a wing disc is wounded, the development of all parts of the organism ceases while the wound in the wing disc is repaired, and then development of the whole organism resumes.

Second, the organism is not specified by its genes, but is a unique outcome of an ontogenetic process that is contingent on the sequence of environments in which it occurs. This can be illustrated by the famous experiments of Jens Clausen, David Keck, and William Heisey on plants from different environments.[5] These experiments took advantage of the fact that in some plants it is easy to clone genetically identical individuals by the simple process of cutting a plant into pieces, each one of which will grow into a new complete individual. A sample of the plant *Achillea millefolium* was taken and each plant was cut into three pieces. One piece was planted at a low elevation, 30 meters above sea level, one at an intermediate elevation in the foothills of the Sierra Nevada mountains at 1,400 meters, and one at a high elevation, 3,050 meters, in the mountains. The three plants that grew from the three pieces of the original plant are then genetic clones of each other developing in three different environments. The result of the experiment for seven different plants is shown in Figure 1.4.

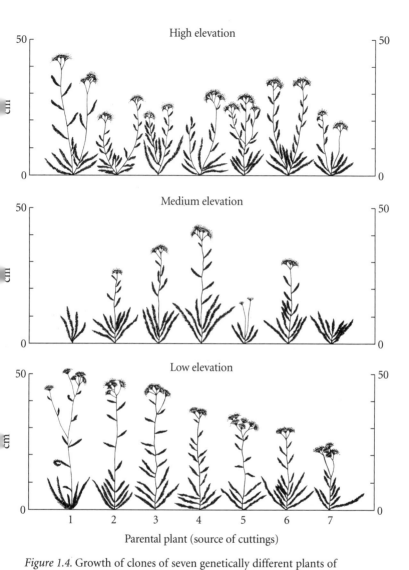

Figure 1.4. Growth of clones of seven genetically different plants of *Achillea* grown at three different elevations.
From *An Introduction to Genetic Analysis* by Suzuki et al., © 1996, 1993, 1989, 1985, 1981, 1976 by W. H. Freeman and Company. Used with permission.

The seven different genetic strains that were sampled are shown horizontally, arranged in order of how well they grew at the lowest elevation. The three plants in a vertical row are the plants that grew from the three cloned pieces from a single plant in the three different environments. We see immediately that it is not possible to predict the order of growth in the medium or high elevation from the order at the lowest elevation. The plant that grew best at the lowest elevation also had the best growth at the highest elevation, but at the medium elevation it was the poorest plant and failed to flower. The second-best-growing plant at high elevation was next to the worst at low elevation and in the middle of the growth range at intermediate elevation. In general, there is no way of predicting the growth order from one environment to another. There is no correlation of growth pattern from one environment to another. It is not possible to ask the question, "Which genotype caused the best growth," without specifying the environment in which the growth occurred. Even averaging over the environments is not very informative. Genotype 5 (average = 25 cm) and genotype 7 (average = 18 cm) grew more poorly on the average over the environments, but the averages of the other five genotypes were indistinguishable (32–33 cm), even though each grew very differently in each environment. It is important to note that Figure 1.4 does not portray an extreme example. The experiments involved many such comparisons, and all showed similar results.

➣ The experiment in Figure 1.4 can be represented in a graphical form that summarizes the results. In Figure 1.5 plant height for each genotype is plotted against the elevation at

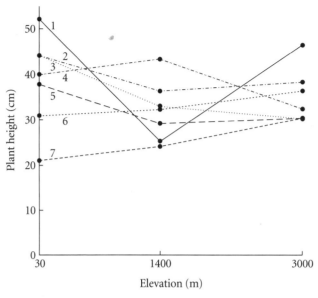

Figure 1.5. A graphical representation of the heights of the seven plants shown in Figure 1.4, at the three different elevations.
From *An Introduction to Genetic Analysis* by Suzuki et al., © 1996, 1993, 1989, 1985, 1981, 1976 by W. H. Freeman and Company. Used with permission.

which it grew. Such graphs, giving the *phenotype* (physical properties) of organisms of a particular genotype as a function of the environment, are called *norms of reaction*. A norm of reaction is the mapping of environment into phenotype that is characteristic of a particular genetic constitution. So a genotype does not specify a unique outcome of development; rather it specifies a norm of reaction, a pattern of different developmental outcomes in different environments. The norms of reaction in Figure 1.5 are typical of what is seen in such experiments. There are occasional genotypes like genotype 7 whose norm of reaction lies below others in all environments. But most genotypes have norms of reaction with complex patterns that cross each other in unpredictable ways. The norm of reaction for genotype 3 decreases monotonically with increasing altitude. Genotype 4 has a maximum at the intermediate altitude while genotype 1 shows a very pronounced minimum at this altitude.

Results like these are not peculiar to *Achillea* or to plants. Figure 1.6 shows a similar experiment in the fruit fly, *Drosophila melanogaster*. It has so far not been possible to clone *Drosophila* in order to make a large number of individuals of identical genotype, but by genetically marking their chromosomes and making specially designed crosses between marked strains it is possible to produce very large numbers of individuals whose genotype is identical for large sections of the genome. Different genetic strains isolated from natural populations of *Drosophila* can then be compared in different environments. Figure 1.6 shows the survivorship from egg to adult of various genotypes taken from a population of *Drosophila* when the immature stages develop at different temperatures. Again we see the characteristic norms of reaction, some decreasing monotonically with increasing temperature, while some have a minimum and some a maximum at an intermediate temperature. There is no genotype with the unconditionally highest survival, and the ordering of survival among genotypes shows no particular pattern from one temperature to another, although generally there is a reduction of survival with increasing temperature. Thus it would be impossible to predict which genotype would be favored by natural selection because of its superior survival, or to explain after the fact why a particular

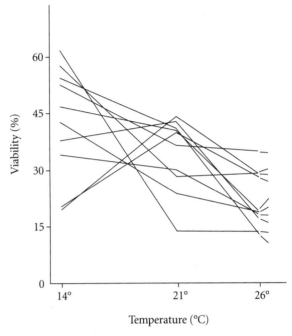

Figure 1.6. The viability of ten different genotypes of *Drosophila* when tested at three different temperatures.

genotype had come to characterize the species, without a specification of the history of temperatures that the species had met in the course of its evolution.

The importance of taking into account the norm of reaction of a genotype is well recognized in plant breeding. New commercial varieties of cultivated plants, for example new maize hybrids, are tested for yield in several years and on farms from different areas in the region where the crop will be grown. Varieties are chosen for release to farmers partly on the basis of their average productivity over years and locations, but also for their uniformity of production over time and space. A hybrid that shows a high average because it is highly superior in a particular year or location, but that otherwise gives a somewhat lower yield than other varieties, will not be selected for release. Seed companies are concerned less with average yield than with reliability of that yield in varying environments, because it is on that basis that farmers will choose the seed to purchase. As a consequence of this policy of plant breeding, there has been an evolution of the norms of reaction of commercial hybrid maize to become flatter and flatter, responding less and less to changes in environment. Figure 1.7 shows a comparison of the norms of reaction of a maize hybrid of the 1940s (Variety 1) and a commercial hybrid from the 1960s (Variety 2), determined in an experiment that compared these different genotypes in a common set of years and locations.[6] In fact, in the best environment the old hybrids were better than the newer ones, but they were more sensitive to different environments and so were replaced by the less environmentally sensitive genotypes.

The actual forms that norms of reaction take also show the error of two more subtle formulations of the relation between genotype and organism which admit some role for environment, but do so incorrectly. One is the notion that genes determine an organism's *capacity,* a limit that may or may not be reached depending on how adequate the environment is. This is the metaphor of the empty bucket. Genes determine the size of the bucket, and environment determines how much is poured into it. If environment is poor, then none of the buckets will have much in it and all genotypes will do poorly, but if the environment is favorable, then the large buckets will be able to

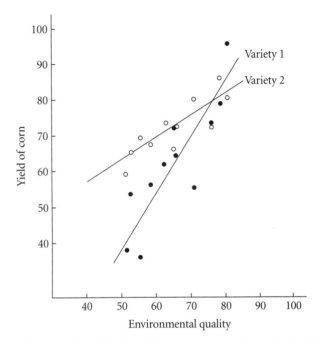

Figure 1.7. The yield of seed from a maize hybrid used in the United States in the 1940s (Variety 1) and a commercial hybrid of the 1960s (Variety 2) when tested in different years and different localities that were rated according to environmental quality.

From *An Introduction to Genetic Analysis* by Suzuki et al., © 1996, 1993, 1989, 1985, 1981, 1976 by W. H. Freeman and Company. Used with permission.

contain a great deal, while the small ones will be filled to their smaller capacity and then overflow.

This capacity metaphor has been widely used in the literature on human IQ. The claim is that IQ will indeed vary over environments, but that in impoverished environments all genotypes will do equally badly, while in enriched environments the genotypes with superior intrinsic capacity will reveal themselves. In this formulation any enrichment of the environment only exaggerates the intrinsic differences that were already immanent in the genotype. Figure 1.8, taken from the famous paper by A. R. Jensen, "How Much Can We Boost IQ and Scholastic Achievement?" makes this argument.[,] However, the norms of reaction shown in the figure are entirely made up by Jensen, and there is no evidence that they represent reality. In a trivial sense every genotype must indeed have a maximum possible metabolic rate, growth rate, activity, or mental acuity in some environment, but, as we have just seen from the actual experimental data on reaction norms, the environment in which that maximum is realized is different for each genotype. Moreover, the ordering of genotypes from "restricted" to "enriched" will change from genotype to genotype. Obviously there will be some environments that will be lethal or severely debilitating for any conceivable genotype, but these are irrelevant to the problem.

There is one sense of "capacity" that is indeed determined by genes. No fruit fly, no matter in what environment it is raised, will be able to write a book about genetics. In a broad and important sense the biology of a species is limited by the possibilities circumscribed by its DNA. As far as we know, genetic differences have no influence on the specific language spoken by a human being, but the possibility of speaking at all depends upon having the right genes. Thus, in answer to a question about why humans and chimpanzees differ in their linguistic abilities, it would be entirely appropriate to say that it is because they have different genes. But the question of the difference between two states is not the same as a question about the

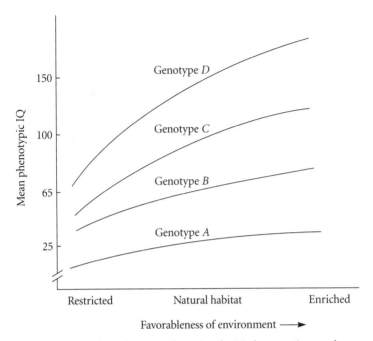

Figure 1.8. Hypothetical norms of reaction for IQ that were invented to illustrate the claim that a trait could be sensitive to environment, yet one genotype would always be superior to another, no matter what the environment.

causation of either of them. Human beings can speak because they have the right genes *and* the right social environment.

Another erroneous understanding of the relation between gene and organism takes yet another step away from determination and says that one genotype has a *tendency* to produce, say, a larger or smaller phenotype than another. In everyday language we say that Bill "tends to be fat" while Ronald "tends to be thin," but it is not clear how this notion is to be used for genotypes and environments. In some environments Bill will be thin and in others, fat. It might mean that on any specific diet Bill will be fatter than Ronald, but if that is its meaning then norms of reaction do not correspond to it, as we have seen. Often the notion of "tendency" carries with it an implicit idea of "normal" conditions or base conditions that hold unless they are disturbed by some outside force. Newton wrote in the *Principia* that bodies tend to stay at rest or in uniform motion "unless compelled to change that state by forces impressed thereon." Thus to make a sensible use of tendency language it must be possible to describe an environment or range of environments in which the phenotype will have the specified form which can be changed only in special circumstances. But in general we do not know how to specify the ideal "normal" environment in which the tendencies of genotypes are to be compared, nor does such an idealized "normal" environment exist any more than does Newton's ideal state without forces.

➤ The view that genotype specifies phenotype is reinforced among geneticists by their long experience of a special class of genotypes that has provided the material for experiments. These are the classic "mutations" in experimental organisms like *Drosophila,* whose norms of reaction are not characteristic of genotypes in general. To be most useful as an experimental tool a mutation should correspond to a phenotypic difference from the "wild type" in every individual that carries the mutation over a broad range of environments. So, the *vestigial wing* mutation or the *white eye* mutation in *Drosophila* can be counted on to cause a shriveled wing or a colorless eye in every individual of that genotype irrespective of the acidity or humidity or temperature of the culture medium in which they developed or of the genetic state of other genes.

Geneticists pretend that these are typical of genetic differences, but what they do not emphasize is that most gene mutations in *Drosophila*, even mutations that geneticists must use in their experiments, are not so well behaved developmentally. The mutation *Curly wing,* for example, widely used for genetic experiments, will result in flies whose wings are observably different from the usual straight wings only if the temperature and humidity of the culture medium are carefully controlled. The majority of known *Drosophila* mutations are like *Curly wing* rather than like *white eyes.* Even mutations that can be reliably distinguished in a wide range of environments are not independent of the milieu in their expression. The *Infrabar* and *Ultrabar* mutations reduce the size of the *Drosophila* eye very considerably and can never be confused with the normal wild type. But the eye size of both mutations as well as of the wild type are responsive to temperature, as shown in Figure 1.9. While the wild type is distinguishable from both mutations at all temperatures, the norms of reaction of the two mutations have opposite temperature trends and cross each other at 15°C.

The genetic determinist view of development presents two alternative schemata for the relation between gene and environment in origin of phenotype. One depicts those basic aspects of the organism that are directly "products" of the genes: its morphology, physiology, cell biology, and innate behavior. Figure 1.10a depicts this schema. There is a basic genetic blueprint that processes different environmental inputs, converting them into organisms whose differences are entirely specified by genetic differences. African Pygmies are extremely short and Dinkas are extremely tall, no matter what their nutrition. The other schema, shown in Figure 1.10b, pertains to those aspects

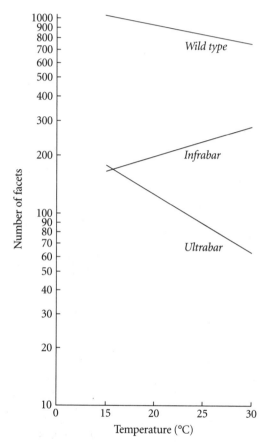

Figure 1.9. The size of the eye, measured by the number of cells (facets), as a function of temperature, for normal wild-type *Drosophila* and two mutant forms, Infrabar and Ultrabar.

From *An Introduction to Genetic Analysis* by Suzuki et al., © 1996, 1993, 1989, 1985, 1981, 1976 by W. H. Freeman and Company. Used with permission.

of the organism that are seen as superficial. In this schema there are basic genetic rules common to all individuals which convert different environmental inputs into different phenotypic outputs. Dinkas and Pygmies speak different languages, learned from their parents, using the same anatomical and neural features.

The schema generated by the norm of reaction that takes account of the developmental interactions between gene and environment is of a very different topology, shown in Figure 1.10c. There are unique interactions between gene and environment such that the ordering of phenotypes has no correspondence to any a priori ordering of genotypes or environments separately. Yet even Figure 1.10c does not capture the complete truth about ontogeny.

Insects have large numbers of sensory bristles arranged in patterned groups on various body parts. Each of these sensory hairs arises from three cells, one forming the bristle, one forming the socket out of which hair grows, and one forming the nerve cell that communicates the bristle motion to the central nervous system. In *Drosophila* one such group is located on the body under the wings. The average number of bristles is the same on the right and left sides, so *Drosophila* is on the average symmetrical. But the number on the left side of an individual is not usually the same as the number on the right side of the same individual. One fly may have nine bristles on the right and five on the left, whereas another fly may have six on the right and eight on the left. This variation is numerically as great as the average difference in bristle number between different individuals, and it is not trivial functionally because the sensory hairs are detectors of the movement of the insect through the air.

What is the source of this fluctuating asymmetry? The cells

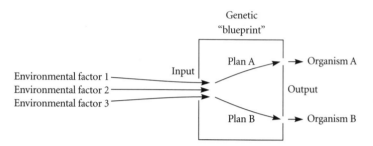

Figure 1.10a. A scheme emphasizing genetic determination of the organism.

From *An Introduction to Genetic Analysis* by Suzuki et al., © 1996, 1993, 1989, 1985, 1981, 1976 by W. H. Freeman and Company. Used with permission.

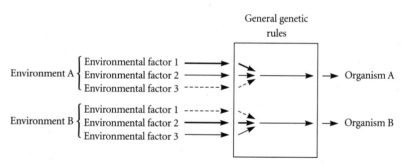

Figure 1.10b. A scheme emphasizing environmental determination of the organism.

From *An Introduction to Genetic Analysis* by Suzuki et al., © 1996, 1993, 1989, 1985, 1981, 1976 by W. H. Freeman and Company. Used with permission.

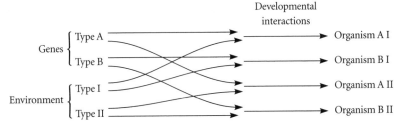

Figure 1.10c. A scheme illustrating the interaction of genes and environment in determining the organism.

From *An Introduction to Genetic Analysis* by Suzuki et al., © 1996, 1993, 1989, 1985, 1981, 1976 by W. H. Freeman and Company. Used with permission.

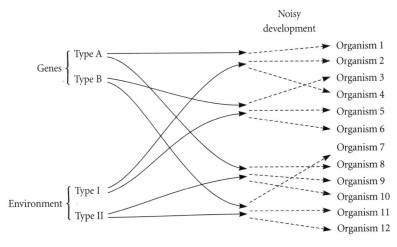

Figure 1.10d. A complete scheme of organismal development showing the interaction of genes and environment, taking into account random developmental noise.

From *An Introduction to Genetic Analysis* by Suzuki et al., © 1996, 1993, 1989, 1985, 1981, 1976 by W. H. Freeman and Company. Used with permission.

on the left and right sides of the fly have the same genes, and it seems ridiculous to say that the developmental environment— the temperature, humidity, oxygen concentration, and so on— was different on the right and left sides of an insect that is two millimeters in length and one millimeter in width and developed its bristles while adhering by its ventral surface to the inside of a glass culture vessel in the laboratory. So the variation is a consequence of neither genetic nor environmental variation. It is *developmental noise,* a consequence of random events within cells at the level of molecular interactions.

Unlike test tubes, cells contain a very small number of many of the molecules that are involved in cell metabolism. The DNA, for example, is contained in exactly two copies in each cell, and many other molecules are not much more numerous. In addition, the molecules are differentially concentrated in different parts of the cell and the cell machinery depends on movement of molecules to meet each other for reactions. The messenger RNA molecule that is the immediate copy of a gene that is being read by the cell must move out of the nucleus and into the cytoplasm in order to take part in the synthesis of proteins. In the cytoplasm it must be inserted into a ribosome, the machine that actually manufactures a protein according to the specification carried by the RNA. This process and all others like it in the cell take time and occupy space and are quite unlike the picture of what happens when billions of small molecules interact with each other by bouncing around in a solution.

The consequence of there being a very small number of chemical units processed by spatially constrained intracellular machines is that there is considerable variation from cell to cell in the rate and number of molecules that are synthesized. This becomes manifest in variation in the time that it takes for cells to divide or to migrate during development. Such variation can be seen in bacterial cells, which are structurally much simpler than the cells of higher organisms. If a large batch of constantly stirred liquid growth medium is inoculated with a single bacterial cell, that cell will divide in, say, sixty-three minutes. But the two daughter cells will not divide simultaneously sixty-three minutes later, and the resulting four cells of that division will not divide again simultaneously. Bacterial cultures do not grow in pulses but continuously, because each cell formed takes a slightly different time to divide. All the cells are growing in exactly the same culture conditions because the culture medium, constantly stirred, is made of high concentrations of small molecules whose local concentration is effectively everywhere the same, and the cells are genetically identical since not enough time has elapsed during the few generations of division to allow many mutations. The cause of their asynchrony is the random uneven distribution of the different kinds of molecules to the daughter cells at cell division. The cells will then need different times to manufacture a new population of necessary molecules before they can divide again.

The same phenomenon occurs in the development of multicellular organisms. The three cells that give rise to a sensory bristle in flies are the result of two divisions of an original precursor cell. To produce an adult bristle, the bristle-forming cells must migrate to the surface of the developing fly, a surface that is progressively hardening. If the division of the original precursor cell into three takes a little too long and the migration of the cluster is delayed, it will not arrive at the hardening surface soon enough to be included as a bristle. Such random processes must underlie a great deal of the variation observed between organisms, including variation of their central nervous systems.

A leading current theory of the development of the brain, the selective theory, is that neurons form random connections by random growth during development. Those connections that are reinforced from external inputs during neural development are stabilized, while the others decay and disappear.[8] But the connections must be randomly formed before they can be stabilized by experience. Such a process of neural development could give rise to differences in cognitive function that were biological and anatomically innate, yet neither genetic nor environmental. I am certain that even if I had studied the violin from the age of five, I could not play a Paganini caprice as Salvatore Accardo does, and Accardo no doubt has neural connections that I lack and has had them since an early age. But it is by no means clear that those anatomical differences between us are genetic. To relate the undoubted existence of random nerve connections to variation in specific characteristics like musical ability would require a major research program. But such a research program will only be carried out if the question is asked in the first place.

The inclusion of developmental noise in the process of development produces the schema shown in Figure 1.10d. The organism is determined neither by its genes nor by its environment nor even by the interaction between them, but bears a significant mark of random processes. The organism does not compute itself from the information in its genes nor even from the information in the genes and the sequence of environments. The metaphor of computation is just a trendy form of Descartes's metaphor of the machine. Like any metaphor, it catches some aspect of the truth but leads us astray if we take it too seriously.

II ORGANISM AND ENVIRONMENT

⋙ The belief that organisms are remarkably well suited to the world in which they live long predates scientific biology. Indeed the extraordinary fit between the properties of living beings and the circumstances of their lives was claimed to be evidence of the power and beneficence of the Divine Creator. How else could we explain that animals living in water have fins for swimming and gills for breathing, while animals that fly through the air have wings and lungs? Together with the question of the origin of diversity among species, this observation of the "fitness" of organisms for their environmental circumstances formed the agenda for Darwin in creating a satisfactory theory of evolution. His solution was that the process of becoming better fitted to the demands of the environment was the same process that would lead to diversity. Those whose anatomy, physiology, and behavior fit the requirements of the environment will have the greatest chance to survive to reproductive age and to produce the most offspring. If those characteristics are heritable, then the next generation will have a higher frequency of individuals with the fitter traits, and eventually the species will consist entirely of the more fit types. But if evolution by natural selection causes organisms to fit better and better into some particular set of environmental circumstances, then groups of organisms separated in time and space will evolve to fit different sets of circumstances as they find them. Organic diversity is then a consequence of the existence of a diverse set of environments to which different species have become fitted by natural selection. The process of that fitting is the process of *adaptation.*

To create such a theory of evolution Darwin had to take a revolutionary step in thinking about organism and environment. Previously there had been no clear demarcation between internal processes and external ones. There was, in the premodern view of nature, no clear separation of living and dead, animate and inanimate. The dead could become alive, statues made of ivory could become living women. Lamarck's theory of evolution assumed the inheritance of acquired characteristics. That is, circumstances outside the organism could be incorporated into the organism in a permanent and heritable fashion through the organism's will. Darwin created a dramatic rupture in this intellectual tradition by alienating the inside from the outside: by making an absolute separation between the internal processes that generate the organism and the external processes, the environment, in which the organism must operate.

In Darwin's theory variation among organisms results from an internal process, what is now known as gene mutation and recombination, that is not responsive to the demands of the environment. The variants that are produced are then tested for acceptability in an environment which has come into being independent of that variation. The process of variation is causally independent of the conditions of selection. The history of the environment, in turn, is a history of geological change, of the impacts of meteors, of the waxing and waning of glacial ages, of the rise and fall of sea level, of shorter-term changes in weather patterns. Other kinds of organisms are also part of the environment of a species, but they appear as given, with histories that are independent. The organism and the environment then interact only through the selective process.

Many metaphors have been invoked for this relation between independent environment and organism. The organism proposes and the environment disposes. The organism makes conjectures and the environment refutes them. In the most popular current form in the technical literature of evolutionary studies, the environment poses problems and the organism throws up random solutions. In such a conceptual structure the metaphor of adaptation is indeed appropriate. Adaptation is literally the process of fitting an object to a preexisting demand. We adapt a key to a lock by filing it to fit the tumblers. When I travel to Europe I carry an adaptor so that my electric toothbrush, designed to work at 110 volts, will function at 220 volts. Organisms adapt to the environment because the external world has acquired its properties independently of the organism, which must adapt or die.

Modern biology not only adheres to Darwin's theory of adaptive evolution by natural selection but bears the marks of the original model of the relation between organism and environment that Darwin imposed. *Fitness* and *adaptive value* are now technical terms for the numerical probability of survival and rate of reproduction of a genotype or phenotype. Thus a population geneticist will say that one genotype has a fitness of 0.78 compared with a fitness of 1.0 of another genotype, although an explanation is rarely offered about the way in which the superior genotype actually "fits" into some environment. As the terms are actually used there may not even be any implication that such a story could be uncovered. So, a genotype whose effect was to interfere with early embryonic development by preventing normal cell division would still be said to have a low "fitness." Nevertheless, the use of the term *fitness* for the numerical force of natural selection generally reinforces the image of the properties of the organism being molded to the specific requirements of the environment. Reciprocally, in ecology, *ecological niche* is a technical term universally used to denote the complex of relationships between a particular species and the outside world. But the use of the metaphor of a niche implies a kind of ecological space with holes in it that are filled by organisms, organisms whose properties give them the right "shape" to fit into the holes.

Together the metaphors of adaptation and ecological niche create an explanation of the observed diversity of organisms. The properties of species map the shape of the underlying external world, just as when we sprinkle iron filings on a sheet of paper lying on a magnet, the filings form a pattern that maps the underlying magnetic field. In a curious sense the study of the organisms is really a study of the shape of the environmental space, the organisms themselves being nothing but the passive medium through which we see the shape of the external world. They are the iron filings of the environmental field. Most evolutionary biologists would reject such a description of their science and would insist that it is the organisms themselves that are the primary objects of interest—yet the structure of adaptive explanation of traits points in the opposite direction.

Adaptive explanations have both a forward and a backward form. In the forward form, usually invoked for extant species, a problem for the organism is described on the basis of knowledge of or supposition about what is important to the organism. Then some anatomical, physiological, or behavioral feature of the species is proposed as the organism's solution to the problem. The backward form, usually used for extinct species known from fossil material, starts with a trait as a solution and searches for the problem that it has solved.

An example that has occupied a great deal of attention is the problem of energy budget. An animal that must forage for its food expends energy in finding nutrition, and its method of search must be such as to provide it with a net positive energy balance after it consumes the food. It is said that during the Second World War underwater swimmers in France tried to supplement their diets by diving in the ocean for fish, but they found that they were slowly losing weight because the energy consumed in the underwater chase at cold temperatures was greater than the caloric value of their catch. Some birds, called "central-place foragers," fly out from the nest, gather food items, and then bring them back to the nest to consume them. If they take the first food item they encounter, it may be so small as not to repay the round trip. But if they take only very large items, such food particles may be so rare that they will expend too much energy in a long search. The solution for maximizing the net caloric intake is to search for food particles that are larger than the average of what is available but not too large, and this optimum bias can be calculated from knowledge of the distribution of food particle sizes in nature.

When birds are actually observed they do take larger-than-average food items, but not as large as they should if they were optimizing their caloric budget. Gordon Orians and Nolan Pearson, who studied this problem, concluded that the birds were compromising between solving the problem of caloric optimization and solving the problem of not staying away from their nest too long in order to protect their nestlings.[1] In this example the investigators began in the forward mode, starting

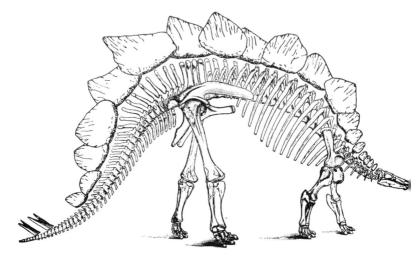

Figure 2.1. A skeleton of the dinosaur *Stegosaurus* showing the row of leaf-like bony plates along the back.
From *Vertebrate Paleontology* by Alfred S. Romer et al., © 1945, 1933 by the University of Chicago. Used with permission.

with the problem of caloric balance and describing the search pattern as the solution, but then switched to the backward mode, starting with the non-optimal foraging strategy as a solution to a problem and then looking for a problem that it might solve.

An example of the backward mode in paleontology is the question of why the dinosaur *Stegosaurus* had a double row of leaf-shaped bony plates along its back (Figure 2.1). To what problem are these plates a solution? Several answers have been given, but it will never be possible to decide definitively among them. In one story the plates are a sexual recognition signal for the species. In another story they make the silhouette of this herbivorous animal larger so that it will discourage attacks by carnivorous predators. Or sometimes it is said that the plates are actually physical defenses against being bitten. Perhaps the most sensible postulate is that they were heat-regulatory cooling fins, an explanation that agrees with their shape and placement on the body and the apparently large number of blood vessels that served them.

Both the forward and backward forms of explanation in these examples seem to make the particular organisms only an excuse for a different project, which is to show how the properties of living beings map the demands of the environment through adaptation. In this view the organism is the object of evolutionary forces, the passive nexus of independent external and internal forces, one generating "problems" at random with respect to the organism, the other generating "solutions" at random with the respect to the environment.

⇒ Darwin's alienation of the outside from the inside was an absolutely essential step in the development of modern biology. Without it, we would still be wallowing in the mire of an obscurantist holism that merged the organic and the inorganic into an unanalyzable whole. But the conditions that are necessary for progress at one stage in history become bars to further progress at another. The time has come when further progress in our understanding of nature requires that we reconsider the relationship between the outside and the inside, between organism and environment. The claim that the forms of heritable variation that arise are not causally dependent on the nature of the world in which organisms find themselves is almost certainly true. There is no credible evidence that acquired characteristics can be inherited or that the process of gene mutation will produce enough of just the right variants at just the right moments to allow species to survive changing environments without natural selection. But the claim that the environment of an organism is causally independent of the organism, and that changes in the environment are autonomous and independent of changes in the species itself, is clearly wrong. It is bad biology, and every ecologist and evolutionary biologist knows that it is bad biology. The metaphor of adaptation, while once an important heuristic for building evolutionary theory, is now an impediment to a real understanding of the evolutionary process and needs to be replaced by another. Although all metaphors are dangerous, the actual process of evolution seems best captured by the process of *construction.*

Just as there can be no organism without an environment, so there can be no environment without an organism. There is a confusion between the correct assertion that there is a physical world outside of an organism that would continue to exist in the absence of the species, and the incorrect claim that environments exist without species. The earth will precess on its axis and produce periodic glacial and interglacial ages, volcanoes will erupt, evaporation from oceans will result in rain and snow, independent of any living beings. But glacial streams, volcanic ash deposits, and pools of water are not environments. They are physical conditions from which environments may be built. An *environment* is something that surrounds or encircles, but for there to be a surrounding there must be something at the center to be surrounded. The environment of an organism is the penumbra of external conditions that are relevant to it because it has effective interactions with those aspects of the outer world.

If the concept of the preexistent ecological niche is to have any concrete reality and any value in the study of nature it must be possible to specify which juxtapositions of physical phenomena would constitute a potential niche and which would not. The concept of an empty ecological niche cannot be made concrete. There is a non-countable infinity of ways in which the physical world can be put together to describe an ecological niche, nearly all of which would seem absurd or arbitrary because we have never seen an organism occupying such a niche. Even a small variation in the description of a known ecological niche leads to possibilities that have never been exploited. There is no animal that flies through the air, lives in trees in nests made of grass, and uses the vast quantity of leafy vegetation available at the tops of the trees for food. Perching birds do not eat leaves. Nor do they eat bark or mushrooms or herbaceous stems or roots. Moreover, there are actual ways of making a living that would seem absurd if we had not observed them. Who could imagine that ants could live by gathering and mulching leaves to make a garden bed in which they would sow the spores of fungi to grow their food? Yet fungus-gardening ants exist.

A practical example of the problem posed by arbitrarily defined ecological niches in the absence of organisms was given by the search for life on Mars. When the first Mars lander was being designed the question arose of how to detect life on the Red Planet. Two basic designs were proposed. The first was a long sticky tongue that would be pushed out of the lander into the Martian dust and then retracted into a microscope. The microscope would transmit pictures back to earth, and these could be examined for objects that looked like some sort of living cell or product of a life form. We might call this the morphological definition of life. The second design, which was eventually adopted, was a tube that would suck up Martian dust into a reaction vessel filled with a growth medium for microbial life. The carbon in the carbohydrate of the medium was radioactively labeled so that the carbon dioxide liberated when cells use a carbohydrate for energy could be detected by a radioactivity counter. We can call this the physiological definition of life.

It would be hard to exaggerate the delirious joy experienced by the scientists who monitored the experiment when, after landing, the machine did indeed send back signals of a rapidly increasing amount of radioactive carbon dioxide in the reaction chamber. Then, suddenly, the production of carbon dioxide ceased, although the machinery was working perfectly. This is a behavior unknown in growing bacterial cultures. As cells begin to exhaust the culture medium in a flask the rate of production of carbon dioxide should cease rising so rapidly, reach a temporary plateau, and then decline continuously as cells start to die from starvation. A sudden shut-down in production cannot be explained. The consensus of scientists working on the problem of extraterrestrial life was that there was no life on Mars and that the original production of carbon dioxide was the result of a breakdown of the culture medium on the surface of the fine dust particles that catalyzed the process. Subsequently such a breakdown of organic compounds on the surface of finely divided clay was reproduced in the laboratory. The problem of the lander was that it presented Martian life with an ecological niche and asked whether that niche was filled on Mars. The designers of the Mars lander believed that ecological niches already exist in the absence of organisms, so that when the organisms evolved on Mars they would come to occupy those empty niches. What could be more reasonable than to suppose that such a basic ecological niche as a carbon source for energy metabolism and some oxygen would be present on Mars? But that ecological niche was assumed to exist by the scientists on the basis of their knowledge of *terrestrial* life.

If niches do not preexist organisms but come into existence as a consequence of the nature of the organisms themselves, then we will not have the faintest idea of what Martian niches will be until we have seen some Martian organisms in action. For all we know, Martian life traps energy by an entirely different mechanism—or perhaps it is just allergic to sugar!

➣ To arrive at a concept of the environment that will be correct and useful for our understanding of past evolution, for our prediction of the future of earthly conditions, and for an efficient search for extraterrestrial life, we need to clarify several facets of the relation between organism and environment. First, organisms determine which elements of the external world are put together to make their environments and what the relations are among the elements that are relevant to them. In my garden there are trees, and grass growing around the trees, and some stones lying here and there on the ground. The grass is part of the environment of a phoebe, a bird that makes its nest out of dried grass, but the stones are not part of its environment. If they disappeared it would make not the slightest difference to the phoebe. But those stones are part of the environment of a thrush, a bird that uses the stones as an anvil to break open snails on which it feeds.[2] There are holes high up in the trees which woodpeckers use for nests, but these holes are not part of the environment of either the phoebe or the thrush. The elements of each bird's environment are determined by the life activities of each species.

The reader should try the experiment of reading or asking an ornithologist for a description of the environment of a bird. The description will be something like this: "The bird eats insects in the summer when they are abundant, but switches to seeds in the fall. It makes a nest of grass and small twigs held together with some mud, built about three meters above the ground in the crotch of a small tree. In the spring and summer it is found as far north as 55 degrees, but in the winter it flies south and is absent above about 40 degrees latitude. In the spring males return first to establish breeding territories, which are later occupied by the returning females." And so on. Every element in this specification of the environment is a description of activities of the bird. As a consequence of the properties of the animal's sense organs, nervous system, metabolism, and shape, there is a spatial and temporal juxtaposition of bits and pieces of the world that produces a surrounding for the organism that is relevant to it.

Nor is this organismal determination of the relevant bits of the world confined to animals whose motor activity makes it possible for them to move from one place to another and to manipulate pieces of the physical world. Insect-pollinated flowers that appear late in the summer are fertilized by an entirely different insect fauna than are early spring flowers. Flowers with long thin corollas are pollinated by hummingbirds and hawk moths that are not part of the environment of flat, open flowers, even though the two kinds of flowers are open side by side at the same moment. Thus fluctuations in the populations of hummingbirds will have a major influence on the pollination success of one of these kinds of flowers but not the other, because these birds are part of the environment of the long thin flowers but not of their immediate spatial and temporal neighbors.

It is, in general, not possible to understand the geographical and temporal distribution of species if the environment is characterized as a property of the physical region, rather than of the space defined by the activities of the organism itself. In animals this may mean that the behavior of the individual allows it to seek out physical conditions in specialized places, so-called microhabitats that are not typical of the more broadly defined region. Only in this way can we understand the seemingly paradoxical behavior of the fruit fly *Drosophila pseudoobscura* in experiments on its humidity preferences. The fly lives in both dry and more humid regions of North America. Researchers expected that when flies were placed in a humidity gradient the flies from the dry region would move toward the drier end of the gradient, while flies from the moister region would prefer the wetter end of the gradient. But the opposite was observed. Flies from the dry regions showed a greater preference for high humidity than those from moderately humid environments. The explanation of these observations lies in the realization that the humidity in which flies are actually living in nature is determined by the microhabitat they choose. No fruit flies can actually live in an environment of very low humidity, because they would dry out and die quite quickly. The flies from the dry part of the range are actually living in small crevices and between leaves where the local humidity is high. Their possibility of living in the generally dry part of the species range depends on their superior ability to seek out the moist microhabitats there. If one wants to know what the environment of an organism is, one must ask the organism.

A second facet of the relation between organism and environment that needs to be clarified is this: organisms not only determine what aspects of the outside world are relevant to them by peculiarities of their shape and metabolism, but they actively construct, in the literal sense of the word, a world around themselves. It is trivial that birds and ants make nests, earthworms live in burrows, and human beings make clothes and houses, but these seem special cases. In fact, all terrestrial organisms, both plants and animals, create shells around themselves that can be observed with simple instrumentation. If motion-picture photographs are taken of, say, a human being, using *schlieren* lenses that can detect differences in the optical density of air, it will be observed that there is a layer of higher-density air surrounding the body, moving slowly upward and off the top of the head. This layer is warm, moist air that is created by the body's metabolic heat and water. It can be seen to surround any metabolizing organism that lives in air, even trees. The consequence is that the individual is not living in the atmosphere as we normally think of it, but in a self-produced atmosphere that insulates it from the outer air. The existence of this layer explains the wind-chill factor, which is a consequence of the insulating layer being stripped away by the wind, leaving the body exposed to the actual surrounding temperature. In normal circumstances it is the warm, moist, self-produced shell that constitutes the immediate space within which the organism is operating, a space that is carried around with the individual just as a snail carries around its shell.

Third, organisms not only determine what is relevant and

create a set of physical relations among the relevant aspects of the outer world, but they are in a constant process of altering their environment. Every species, not only *Homo sapiens*, is in the process of destroying its own environment by using resources that are in short supply and transforming them into a form that cannot be used again by the individuals of the species. Food is turned into poisonous waste products by every metabolizing cell. Plants suck up water from the soil and transpire it into the air. Although water is returned to the soil, its local rate of replenishment is essentially independent of its local rate of extraction, so that plants in a particular place are creating their own drought.

But every act of consumption is also an act of production. That is, living systems are the transformers of materials, taking in matter and energy in one form and passing it out in another that will be a resource for consumption for another species. The waste products produced by the consumption of food by one species are, in turn, the food of other species. The excrement of large herbivores becomes the sustenance of beetles. The carbon dioxide produced by animals is the raw material for plant photosynthesis. Thus all organisms alter not only their own environments but also the environments of other species in ways that may be essential to the life of those other organisms. The simplistic and incorrect understanding of Darwinism, that nature is "red in tooth and claw," that all organisms are in a constant state of competition, that one must eat or be eaten, misses entirely this productive side of life processes. The satirist Mort Sahl used to say, "Remember that no matter how selfish, how cruel, how unfeeling you have been today, every time you take a breath, you make a flower happy."

Nor is this productive alteration only the effect of one species on another. It is very well known that the roots of legumes contain nodules of bacteria that turn the gaseous nitrogen of the air into fixed soluble nitrates in the soil. These nitrates are then taken up by the roots to nourish the same plant that produced them. But root systems do more. They physically condition the soil by breaking it up, changing the form, size, and composition of the soil particles in such a way as to make further root development easier. At the same time they excrete humic acids into the soil, which encourage the formation of mycorrhizal associations with soil fungi. These symbiotic associations in which the fungi penetrate the plant root tissues are important mechanisms of plant nutrition.

The concept of "alteration" of the environment does not capture entirely the way in which organisms mold their immediate local conditions. The sunlight, temperature, humidity, and wind velocity recorded in government records and reported in newspapers are determined by weather stations at the tops of buildings or mountains or in open fields. But these are not the conditions that exist in fields of cultivated plants like maize or in forests. The light intensity, temperature, humidity, air movement, and gaseous composition of the atmosphere in a densely cultivated field or a forest all vary with height from the ground. The microclimate near the soil surface is quite different from that between two lower leaves of a maize plant, which is again quite different from the microclimate for leaves near the growing top of the plants. The zones change as the plant grows taller and as the leaves grow longer and touch the leaves of neighboring plants. These microclimatic variations play an extremely important role in growth and production because it is the intensity of solar radiation and the carbon dioxide concentration at the surface of the leaves that determine the rate of photosynthesis and thus the growth rate and productivity of the maize plant. So the rate of growth determines the microenvironment, which determines the rate of growth.

Not only the rate of growth but the exact morphological pattern of leaves is an important variable. The spacing of leaves along the stem and their position around the stem, the shape of each leaf, its angle of repose against the stem, the hairiness of its surface determine how much light, moisture, and carbon dioxide reach the leaves and how rapidly oxygen produced by photosynthesis is carried away. And all of these affect the plant in a way that is characteristic of the pattern of development.

The practical consequence of all this complexity is seen in the science of plant engineering. In an attempt to increase the productivity of crops, plant engineers make detailed measurements of microclimate around the plant and then redesign the pattern of leaves to increase the light falling on the photosynthetic surfaces and the available carbon dioxide. But when these redesigned plants, produced by selective breeding, are tested it turns out that the microclimatic conditions for which they were designed have now changed as a consequence of the new design. So the process must be carried out again, and again the redesign changes the conditions. The plant engineers are chasing not only a moving target but a target whose motion is impelled by their own activities. As we will see, this process is a model for a more realistic understanding of evolution by natural selection.

➤ The notion that organisms are chasing a moving target during their evolution has a wide currency. In 1973, Leigh Van Valen pointed out a seeming paradox in evolutionary theory.[3]

If organisms are constantly adapting to the outer world, then as evolution goes on species should be better and better able to survive the rigors of the environment and so they should endure for longer and longer periods. But when Van Valen examined the fossil record he found that the time between first appearance and disappearance of forms has not grown longer over evolutionary time. His conclusion was that the environment is constantly changing so that adaptation to yesterday's environment does not improve the chance of survival tomorrow. He called this the "Red Queen Hypothesis" after the chess queen in *Through the Looking Glass* who found that she had to keep running just to stay in the same place because the ground was moving under her feet. The Red Queen, however, is not the same as a constructionist view of the organism and its environment. Even if the external world is changing in ways that are completely independent of the organisms, organisms will still have to run to keep up. The constructionist view is that the world is changing *because* the organisms are changing. The Red Queen's running only makes the problem worse.

Another consequence of the organism's reconstruction of its own environment is a struggle between generations. In rural regions of the northeastern United States, such as Vermont or northern New York State, the maximum human population density was reached around 1850. At that time nearly the entire land surface was occupied by farms. Farming in these regions with thin soils and a short growing season was difficult, so that when government policy encouraged the opening of new lands in the Middle West, where the growing conditions for field crops were ideal, there was a massive exodus of population. As a result, much of the farmland of the Northeast returned naturally to forest. In New England, this return begins with the growth of herbaceous plants and low woody shrubs, but these are soon succeeded by white pine, which forms dense pure stands. In the early part of the twentieth century these stands of white pine became mature and were heavily harvested for timber and paper. The owners of the land, mostly large paper companies, then attempted to regrow the pine forests, but consistently failed because there was a natural growth of hardwood trees that crowded out the pine seedlings.

This same process occurs naturally. In New England, as white pines mature and begin to die or are blown down, they are not replaced by their own seedlings, but by hardwoods. The failure of the second generation of pines is partly a consequence of the sensitivity of pine seedlings to low light intensities as compared to the tolerance of hardwood seedlings. In the cut forests, as in the naturally maturing ones, seedlings of hardwoods are able to live in the shade of the mature pines, and as soon as mature trees disappear these suppressed hardwoods grow quickly, overshadowing any pine seedlings. Adult pines create an environment of deep shade that is hostile to the growth of their own offspring. The conditions that gave rise to the pine forests were changed by those forests so that they could not reproduce themselves. This developmental history of the forest, beginning with abandoned open fields, followed by herbs and shrubs, followed by white pine, followed by hardwoods, is the biological *locus classicus* for the phenomenon of weed plant succession studied by ecologists.

Weeds are precisely those species which can grow only in disturbed conditions, roadsides, gardens, burned areas, harvested forests, and which, having grown, change the conditions of the area so that they cannot produce a second generation. Such species can continue in existence only because disturbances are constantly occurring in one place or another and the weed species have mechanisms for wide random dispersal. The phenomenon of the weed is a manifestation of a general principle of historical development of any system: that the conditions which make possible the coming into being of a state of the system are abolished by that state.

The fourth aspect of the construction of environment is that organisms modulate the statistical properties of external conditions as those conditions became part of their environment. Living systems can perform both time averaging and rate detection. That is, like mathematicians, they can perform integration and differentiation.

In temporally and spatially varying conditions, organisms need to be able to smooth out the effect of the variation over their lifetimes. External sources of energy are not available at all times. Plants photosynthesize during the day but not at night, during the spring and summer but not during the winter. Desert plants may be able to acquire enough water to germinate and grow only in one year out of five, when there is an occasional rainstorm. In temperate regions there is no production of food for herbivorous animals during half the year. It must be possible to average out these periodic fluctuations, storing materials or energy from productive periods, which can then be consumed when there is no production. That is, the relevant aspects of the environment must appear relatively constant to the organism's physiology, even though there are fluctuations in the external world that produces the materials from which the organism's environment is constructed.

Animals have certain tissues in which fat is laid down during periods of high nutrition and from which it can then be metabolized during periods of dearth. A special form of this fat storage is the yolk of eggs in insects, reptiles, and birds, which provides energy during the period of development when the immature animal cannot feed independently. This is carried a step further in those insects whose adult form is developed inside a pupal case or cocoon. The larva or caterpillar is simply a feeding machine that stores up fat by eating voraciously. The pupa then forms, all of the larval structures are destroyed except the embryonic tissue, and the adult is built anew using energy from the stored fat. Plants store energy in underground tubers or in the carbohydrates and proteins of seeds so that the next year's growth or the next generation can begin from internally stored energy.

In a recursive fashion, one species can use the time integration of energy of another one for its own purposes. Oak trees store photosynthetic energy by making acorns, and squirrels store energy by appropriating the acorns and storing them. Human culture has created special devices for such recursive appropriation. Ritual feasts and gift-giving ceremonies like the *potlatch* of Pacific Coast Indians occur during times of abundant resources as a hedge against bad times. Maize seeds are stored energy, which is then fed to pigs that store the energy in fat, which are then slaughtered and smoked or frozen to store energy for consumers, who pay roughly the same price for pork at all seasons of the year because commodity markets (socalled futures) even out the price fluctuations between productive and nonproductive seasons. Money is a time integrator of resource availability through savings, investments, and loans.

It is not only energy that is integrated over time, but also signals from the physical world. Many plants flower when the total number of degree-days accumulated above a certain temperature has reached a critical threshold. The onset of and release from winter hibernation in mammals like bears occur when the accumulated level of certain compounds in the blood reaches a critical level. A few cloudy or cold days will not cause birds to migrate southward or leaves to fall from deciduous trees. In all these cases there is some physical transduction of external signals into stored chemical information that serves as a trigger for physiological and behavioral changes at critical thresholds.

Organisms also differentiate with respect to space and time so that they can detect and react to rates of change of external conditions. From the standpoint of the life processes of some organisms, it is the rate of change, rather than the absolute level of a given physical factor, that is part of the environment and that has been incorporated into the environment by the nature of the organism. It is common among invertebrate animals to alternate between sexual and asexual forms of reproduction. In the case of parasites the signal for the change from asexual to sexual is the shift from an original host species (the primary host) to a secondary host that has been acquired more recently in the parasite's evolution. The reasonable explanation of the change in reproduction is that certain genotypes have been selected during the course of the parasite's evolution to make it survive well in its original host, but that new genotypes are required when it shifts from one host to another. Asexual reproduction maintains the original genotypes unchanged, while sexual recombination produces a wide array of new genotypes, some of which may be well suited to the newer host.

In the case just described there is a direct correspondence between the reproductive form and the identity of the host; it illustrates how a species may alter its reproductive method in response to a "worse" environment. A more interesting situation occurs where the direction of the change is irrelevant. In *Cladocera,* small fresh-water arthropods, reproduction remains asexual as long as conditions of temperature, oxygen dissolved in the water, food availability, and degree of crowding remain constant. Then, if a sudden change in these conditions occurs, whether it increases or decreases food, temperature, oxygen, or space, the *Cladocera* switch to sexual reproduction. It is not the level of these factors but a change in level that is the trigger for sex. The organisms are detecting a rate of change of an input, not its absolute value. They are performing mathematical differentiation.

Finally, organisms determine by their biology the actual physical nature of signals from the outside. They transduce one physical signal into quite a different one, and it is the result of the transduction that is perceived by the organism's functions as an environmental variable. For a mammal, when the temperature of the air rises, the increased thermal agitation of the molecules does not result in a matched increase in thermal agitation of molecules inside the animal. The smallest initial change in internal temperature is converted by the hypothalamus to an endocrine signal which results in a large number of internal chemical and neural and anatomical changes such as concentrations of hormones, of blood sugar, of breathing rate, of the chemical activity of sweat glands, of the contraction of muscle fibers in the skin.

This transduction of one kind of signal into another is a consequence of the internal biology of the species, and different species are different in this respect. So, a rattlesnake subjected to the same increase in thermal agitation of the air molecules will have the thermal agitation of its internal molecules increased considerably, with the result that it will actively change its environment by crawling under a rock or a bush. If I am walking in the desert and disturb the rattlesnake, signals will reach me in the form of photons falling on my retina and compressions of the air falling on my eardrums. These physical signals will immediately be transduced by my physiology into a large increase in the concentration of adrenaline in my blood stream, resulting in sweating, increases in blood pressure, heart rate, and breathing rate, and an impulse to flee. If, in contrast, the same photons and air compressions indicating the presence of a rattlesnake reached another rattlesnake, a very different internal chemical change would occur, perhaps leading to copulation.

These are simple and obvious examples of the generality that it is the biology, indeed the genes, of an organism that determines its effective environment, by establishing the way in which external physical signals become incorporated into its reactions. The common external phenomena of the physical and biotic world pass through a transforming filter created by the peculiar biology of each species, and it is the output of this transformation that reaches the organism and is relevant to it. Plato's metaphor of the cave is appropriate here. Whatever the autonomous processes of the outer world may be, they cannot be perceived by the organism. Its life is determined by the shadows on the wall, passed through a transforming medium of its own creation.

➣ It may be objected that such a view of the relation of organism and outside world ignores some universal physical forces and principles from which no living being can escape. After all, organisms did not invent the law of gravitation. There are indeed physical relations within which the organism must construct its environment, but the physical constraints that are "universal" often turn out to be effective only for certain classes of organisms and not for others. Different species live in different domains of physical forces at a macroscopic level.

The universal law of gravitation is an example. Whether or not gravitation is an *effective* factor in the environment of an organism depends upon the organism's size. Animals of a medium or large size, such as vertebrates, are anatomically constructed under the constraint of gravity. So, large dinosaurs, like elephants, needed bones with a very high ratio of thickness to length. The strength of bones increases with the cross-sectional area and therefore proportionally to the square of linear dimensions, but the weight that the bones must support increases as the cube of linear dimensions. In contrast, bacteria living in a liquid medium are not effectively subject to gravity, which is a negligible force for objects of such a small size floating in a liquid medium. But the difference in size between elephants and bacteria is coded in their genes, so, in this sense, the organisms' genes have determined whether gravitation is or is not relevant to them. Indeed, the largest dinosaurs mitigated the effects of gravity by living partly submerged in water, another consequence of the genes they carried.

Bacteria, although ignoring gravity in their construction because they are so small, are strongly subject to a different "universal" physical force because of their size. If one observes bacteria in liquid culture through a microscope one can see that they are buffeted about by the thermal agitation of the molecules in the culture medium, the force producing Brownian motion. We, in contrast, are not constantly knocked back and forth by the molecules of the air, because we are too large for

Brownian motion to affect us. Differences of size and of the medium in which organisms live are of overwhelming importance in determining the organisms' entire set of environmental relations, but these factors are a consequence of the internal biology of the species.

It is also necessary to realize that life as a whole is evolving in external conditions that are the consequence of the biological activities of that life. Earth's atmosphere at present contains about 20 percent oxygen and 0.03 percent carbon dioxide, and these amounts set severe constraints on the evolution of species. But the modern composition of the air is a direct consequence of the biological activities of ancient life. The pre-biotic atmosphere, like the atmosphere of other planets, contained almost no free oxygen because that element is so chemically reactive that it existed only in chemical combination. In fact most of it was in the forms of carbon dioxide, which was in high concentration in the atmosphere, and water. The oxygen now present in the air was put there by the photosynthesis of green plants. Those same plants trapped the carbon dioxide in solid form, either in the calcium carbonate of limestone, laid down by algae, or in fossil fuels. The external physical conditions that constrain the evolution of modern organisms were constructed by their ancestors.

➣ Are there any circumstances in which it can be said that organisms "adapt" to an externally imposed environment rather than "constructing" it by their life activities? Farmers spray their fields with insecticides that present an external challenge to insects. Major volcanic eruptions can fill the atmosphere of the entire earth with enough dust to affect the growth of plants everywhere, although only for short periods. If, however, it is true that large bodies from space have crashed into the earth occasionally, the perturbations of the atmosphere they created would have affected all of life over a longer time and may have caused the extinction of vast numbers of species. There arc progressive changes in the angle of the earth's axis of rotation with respect to the sun, changes in the eccentricity of the earth's orbit and the movements of continental land masses. One consequence of these is a periodic cooling and warming of the entire surface of the earth. Thus glacial ages come and go periodically and organisms must adapt to them.

Yet even in these cases the biology enters into the determination of the external challenge. As insects adapt to insecticides by becoming more resistant, they induce the farmer to increase the frequency of spraying and to change the chemical. Thus they construct their own hostile environment. Even for major global change, the relevance of a challenge to a species is in part a consequence of the biology of that species. Terrestrial plants and the large herbivores that depend on them for life are far more affected by volcanic eruptions and meteor impacts than are aquatic species. The mass extinctions of the past did not strike species at random, but according to their biology. The picture of evolution that postulates an autonomous external world of "niches" into which organisms must fit by adaptation misses what is most characteristic of the history of life.

There is an immediate political consequence of the appreciation that all organisms construct their own environments and that there are no environments without organisms. The growing environmentalist movement to prevent alterations in the natural world that will be, at best, unpleasant and, at worst, catastrophic for human existence cannot proceed rationally under the false slogan "Save the Environment." "The environment" does not exist to be saved. The world inhabited by living organisms is constantly being changed and reconstructed by the activities of *all* of those organisms, not just by human activity. Neither can the movement proceed under the banner of "Stop Extinctions!" Of all species that have ever existed, 99.99 percent are extinct, and all species that currently exist will one day be extinct. Indeed all life on earth will one day be extinct, if for no other reason than that the sun will eventually expand and burn up the earth, about two billion years from now. As life originated more than two billion years ago, we can say with confidence that life on earth is half over. There is no evidence that living organisms have in their evolution become somehow better adapted to the world. Although the average time from origination to extinction of species has fluctuated from era to era because of glacial ages, the drifting apart of continents, and occasional collisions with meteors, it has not shown any long-term tendency to increase. Nor is there any factual basis for claims that species are in some sort of harmony or balance with each other or with the external world. We cannot prevent environmental change or species extinction. It will take all the political force that can be marshaled just to influence the direction and rate of change of the natural world. What we can do is to try to affect the rate of extinction and direction of environmental change in such a way as to make a decent life for human beings possible. What we cannot do is to keep things as they are.

III PARTS AND WHOLES, CAUSES AND EFFECTS

⯈ The previous two chapters dealt with two aspects of a common theme. How is the natural world of objects and processes to be broken up in such a way as to provide a proper understanding of the history and operation of natural phenomena? The problem of how to parse the world into appropriate bits and pieces is a consequence of the analytic tradition that modern science has inherited from the seventeenth century. If the animal is like a machine, as Descartes claimed in Part V of the *Discourse on Method,* then it is made up of clearly distinguishable bits and pieces, each of which has a determined causal relation to the movement of other bits and pieces.

But Descartes's machine model is not only a description of how the world operates but also a manifesto for how to study natural phenomena. If I wish to study an animal as a machine, I commit myself to behaving as if the animal can be broken down into pieces whose identity as pieces is unproblematic and which have a clear chain of causal connections with each other in producing the properties of the whole. If, wishing to study the operations of a mechanical clock, I open it, I will see a collection of gears, levers, and springs whose status as the parts of the clock are never in doubt. There is no question about where one gear starts and another ends, nor that these immediately perceived separate pieces are the elements whose functional relations need to be specified in any explanation of the operations of the clock as a whole. Furthermore, by removing, altering, or interfering with the operation of each gear and lever separately, I can analyze the chain of forces driving and regulating the clock. The entire explanation will be framed in terms of how the spring transmits a motive force to one gear which transmits forces to a second gear and so on to a third, all at rates determined by the number of teeth in each gear and limited by the escapement mechanism. There is a clear chain of causes and effects among the predefined physical elements of the machine.

Such an analytic mode of understanding and study of biological systems, appropriate to a machine, is implied in the very word *organism,* first used in the eighteenth century. The analogy is between the living body and the musical instrument composed of separate parts that work together to produce a variety of final functions. This was a radical departure from the holistic pre-Enlightenment view of natural systems as indissoluble wholes that could not be understood by being taken apart into bits and pieces, a view echoed in Alexander Pope's simile:

> Like following life thro' creatures you dissect,
> You lose it in the moment you detect.

Over the last three hundred years the analytic model has been immensely successful in explaining nature in such a way as to allow us to manipulate and predict it. It seems abundantly clear to us now that the holistic view of the world obstructs any possibility of a practical understanding of natural phenomena. But the success of the clock model, in contrast to the failure of obscurantist holism, has led to an overly simplified view of the relations of parts to wholes and of causes to effects. Part of the success of naive reductionism and simplistic analysis comes from the opportunistic nature of scientific work. Scientists pursue precisely those problems that yield to their methods, like a medieval army that besieges cities for a period, subduing those whose defenses are weak, but leaving behind, still unconquered, islands of resistance. Science as we practice it solves those problems for which its methods and concepts are adequate, and successful scientists soon learn to pose only those problems that are likely to be solved. Pointing to their undoubted successes in dealing with the relatively easy problems, they then assure us that eventually the same methods will triumph over the harder ones. If the determination of DNA sequence has solved the problem of how information about protein structure is stored in the cell, then surely the determination of the structure of some molecules, perhaps even DNA itself, will solve the problem of how information about social structure is stored in the brain.

Of course, not *all* the information about protein structure is stored in the DNA sequence, because the folding of polypeptides into proteins is not completely specified by their amino acid sequences. That fact is conveniently ignored, because under the physiological conditions of normal cells the folding is unique. When cells are abnormal, however, or when genes from humans are put into microorganisms growing in liquid culture, different outcomes of the folding process may occur, because the "correct" final structure of a protein depends on the formation of correct folding intermediates that will not occur if the external conditions are not appropriate. We do not, in fact, know what the rules of protein folding are, so no one has ever succeeded in writing a computer program that will take the sequence of amino acids in a polypeptide and predict the folding of the molecule. Even programs that attempt very crude characterizations of the folding of regions of proteins into major structural classes like alpha-helices and beta-sheets are not more than about 75 percent accurate. The difficulty is that a protein is not a string of amino acids, although it may be built up from them. It is a unique molecule with unique vibrational and three-dimensional steric properties that change during the process of partial folding. As a result the process of minimizing free energy during folding is chasing a moving target. Molecular biologists do not usually call attention to this ignorance about the determination of protein structure but instead repeat the mantra that DNA makes proteins.

➣ Despite the extraordinary successes of analytic and reductionist biology, the most interesting questions remain: the problems of mind and shape. What are the neurophysiology and neuroanatomy of specific stored memories? Are the same memories in the same "places" at different times? Even the simplest computer changes the location of information in memory as new information is added. And what about conscious attention? As I write this chapter I think at one moment of the sentence I am writing, but then I wonder which sonata my wife will practice next, and then I recall the work done by the plumber today, and then I return my attention to the manuscript. What determines which of the information stored in my brain is in my "mind" at each moment? The difficulty of the problem is not that we lack some crucial bit of knowledge, but that we do not know how to frame the questions. Trapped by the machine model, we have passed through a succession of fashionable metaphors in different technological eras. Once the brain was a telephone switchboard, then it was a hologram, then it was an elementary digital computer, then a parallel processing computer, and now it is a distributed processing computer.

Our ignorance of the generation of organic shape also remains profound, despite the progress made by molecular studies of development. What developmental genetics has done is to substitute a question that it can answer for one that it cannot, but without an explicit acknowledgment of the switch. The original question was why the objects that grow on the sides of my head have the shape and structure of ears rather than of feet and why they look like a human's ears and not like an elephant's ears. The question answered by developmental genetics is which genes are being read by the cells at the front end of an embryo and which at the back end. But which genes are read is not an answer to the problem of shape, a question that must eventually be answered in terms of the determination of the genesis of internal cellular structures (another problem of shape), of the plane of cell division, of the number of cell divisions, of the sliding and folding of sheets of tissue and, above all, of what is called "positional information," the determining influence on these processes of the location on the body where they are taking place.

The difficulty of applying the simple machine model to the study of organisms arises from three sources. Organisms are intermediate in size, they are internally heterogeneous in ways that are relevant to their functions, and they enter into complex causal relations with other heterogeneous systems. There are several consequences of these features that make the simple machine model inappropriate as a mode of understanding or of analysis. First, there is not a single and obvious way to partition an organism into "organs" that are appropriate for the causal analysis of different functions. Second, the organism is the nexus of a very large number of weakly determining forces, no one of which is dominant. Third, the separation of causes and effects becomes problematical. Finally, organic processes have an historical contingency that prevents universal explanations.

The problem of how to determine the appropriate ways of cutting up an organism and its functions lies at the base of many of the most contentious disagreements in biology. There is at the present time a serious split between molecular biologists, who insist that the ultimate explanation of living organisms can be obtained only through a description of the structure and chemical properties of their molecules, and organismic biologists, who claim that it is the whole organism that matters, especially for an understanding of the evolution of life. It is the whole organism that lives or dies, that reproduces more or less, and therefore it is the whole organism that is the object of natural selection.

But this opposition between the individual molecule and the whole organism as the appropriate level of observation and explanation is a false one. It is true that molecular biology, in its most extreme reductionist form, does seem to claim that the structure of an isolated molecule has immense explanatory power. That is the significance of the notion that DNA is "self-replicating." But in the actual study of molecular biology the real object of investigation is the interaction of a molecule with others, for example in the elucidation of the mechanism of the synthesis of new DNA strands using old ones as templates or the synthesis of proteins using information contained in DNA sequences. To carry out the program of molecular studies of cell function or development it is necessary to map out the pathways of causal connection between molecules, because there is no collection of molecules that can be known *a priori* to form a relevant functional unit. In a sense, molecular approaches to biology are attempts to build up the units of "natural" causal relations from individual elements.

Reciprocally, organismic biologists never actually use the whole organism as their unit of study and explanation, but always anatomize it in some way, without offering a clear criterion for how this is to be done. Evolutionists try to explain the evolution of the hand or the brain or the circulatory system or leaf shape or flower morphology. In doing so they break the organism down into pieces by some intuitive process that tells us more about the psychology of human perception than about the actual connections between parts of animals and plants. Under what circumstances is the hand the unit of evolution and function rather than the finger or one joint of the finger?

An example of the difficulties that arise from the arbitrary anatomizing of an organism is the problem of the evolution of the human chin. The anatomy of humans is generally an example of the phenomenon of *neotenic* development. That is, the adult human resembles the fetal ape more than it resembles the adult ape. During the later fetal development of the ape parts of the face and skull, for example brow ridges and the sagittal crest on the top of the skull, grow differentially larger, while humans appear to be born with the juvenile ape characters. An exception to this rule is the chin. Fetal apes and early fetal humans have a receding chin, but as the human fetus develops further the chin becomes more prominent. There have been many speculations about why natural selection has favored a protruding chin in humans, making it an exception to the rule of neoteny. The answer seems to be, however, not that the chin is specially adaptive but that it does not exist!

The feature of the face that we identify as a chin is the consequence of the growth of two independent bones, the dentary bone, in which the teeth of the lower jaw are inserted and the mandibular, which forms the jawbone proper. In human evolution both of these bones have been growing shorter relative to the rest of the skull, and both show neoteny. However, the dentary has been receding in evolution faster than the mandibular, with the result that a distinct protrusion, the "chin," now appears. This object is not an integrated unit of either development or function but an accidental shape to which a name has been given in ordinary perception and which has then become an object of scientific study. The error of arbitrary aggregation is deeply embedded in much of adaptive explanation, especially in attempts to give evolutionary explanations of human social behavior. Thus sociobiologists provide adaptive stories about natural selection for a universal human tendency to form religions, although most cultures, including the classical Greeks, have no separate social function (or word) that corresponds to the modern Western category "religion."

It might be that the appropriate units of study are the units of function, but this viewpoint only brings the ambiguities into sharper focus. One fundamental difficulty in finding the "natural" sutures between parts of an organism is that there are functions at different levels of aggregation. The circulation of the blood serves the vital function of cell respiration by bringing oxygen and removing waste products, so the heart seems a natural anatomical unit. But the contraction of muscles serves the function of making the heart beat, so the structure of muscle cells and their pattern of the innervation is an appropriate level of study. But the shortening of individual muscle fibers serves the function of muscle contraction, and this depends on the chemistry of the proteins actin and myosin. There is a hierarchical cascade of functions that serve other functions above them, and no one of these levels is uniquely correct for the analysis of either the operation or the evolutionary history of the organism. The other problem of function is that in addition to the vertical hierarchy of functions there is a horizontal multiplicity of functional pathways that define parts according to different topologies. Bones serve the function of providing rigidity to the body and attachments for muscles. But they also are the sites for the storage of calcium, and the bone marrow is the tissue within which new red blood cells are produced. Depending on the causal pathway of interest, "bones" are either macroscopic structural elements or collections of cells that secrete calcium or embryonic tissue of the circulatory system.

The functional approach to the definition of parts clarifies the actual process of definition. To be "parts" things must be parts of *something*. That is, there are no parts unless there is a whole of which they are the pieces. For biological systems, because of the hierarchy of functions and because of the multiple intersecting causal pathways, the determination of parts is made only after the appropriate "whole" is defined. Unlike simple physical systems like the planets circling the sun, that whole is not defined by the space it occupies but by function and by the causal pathways that serve that function. Aristotle thought that the function of the brain was to cool the blood, and indeed the very extensive network of blood vessels on the surface of the brain does play an important role in radiating heat from the body. From the standpoint of thermoregulation, the division of the brain into a cerebellum and a cerebrum with temporal, parietal, and frontal lobes is meaningless. Only when the brain is seen as serving sensory, motor, and cognitive functions do the usual neuroanatomical divisions become proper parts of that whole.

⮞ The reductionist world view that dominates our investigation of nature ordinarily leads that investigation to proceed in two stages. To begin there is a downward analytic process that breaks the whole into its constituent parts, which is then followed by a synthetic phase in which the causal pathways among the parts are discovered. That method of investigation works to the extent that the definition of the whole is clear and there is an obvious anatomy of the system. The solar system is the collection of the sun, the planets, and their moons. The entire dynamics of that system can be expressed in terms of the masses, distances, and velocities of those spatially defined objects and the forces operating on them. Biological investigation, in contrast, often begins with an upward synthetic process, in which objects and phenomena are thought to be parts, but the whole of which they are parts is as yet undetermined. As in a play by Pirandello, they are characters in search of an author.

The Human Genome Project, whose goal is to sequence all of the DNA of a human genome (actually a composite of a number of different humans), is precisely of this form. The first step will be to describe the complete ordered sequence of A, T, G, and C that makes up the three thousand million nucleotides of the DNA. Next, this sequence must be broken into pieces of various lengths that correspond to functional units, the genes and their regulatory elements. There are signals internal to the sequence that provide guesses about the boundaries of the part of a gene that is translated into protein, but these are only guesses and can only be confirmed when a protein is actually found in the organism. Moreover, it is virtually impossible to tell where the boundaries of regulatory sequences are, and each case must be investigated by a laborious process of making changes in the DNA and finding the physiological or developmental consequences of those changes. Thus it is impossible to know how to break up a DNA sequence into genes before we know how the cell reads different parts of the DNA in the process of making protein. But this is only the first step. Even after we identify all the genes as functional units in the production of proteins, we will not know the function of those proteins. We will then not know how to assemble the collection of genes and their proteins into functional subsystems with pathways of causal connections. Rather, we will be in the situation of the paleontologist who knows that *Stegosaurus* had large bony plates along its back but must ask the question, "What are they for?"

In biology, this "what for" question is not the same as it is in the analysis of the parts of a motorcar or a clock. In the latter case all the functions are known in advance and it is certain that all the internal parts serve one or another of these functions. In the case of the organism there are, of course, general functions such as motion, respiration, and reproduction that are common, but there are many particular functions, peculiar to different life forms, that cannot be known in advance. In addition, it is by no means true that every part serves a function. Many features of organisms are the epiphenomenal consequences of developmental changes or the functionless leftovers from remote ancestors. Only a quasi-religious commitment to the belief that everything in the world has a purpose would lead us to provide a functional explanation for fingerprint ridges or eyebrows or the patches of hair on men's chests. In biology we cannot escape from the dialectical relation between parts and wholes. Before we can recognize meaningful parts we must define the functional whole of which they are the constituents. We will then recognize quite different ways of breaking up the organism depending upon what we are trying to explain. The hand is the appropriate unit for investigation if we are concerned with the physical act of holding, but the hand and eye together are an irreducible unit for understanding how we come to seize the object that is held. .

➥ The issue of how to break down the organism into separate effectively independent systems that are explanatory is one that exists even at very low levels of description. An example is the question of whether it is possible to describe the evolution of a single gene on the basis of the effects of different states of that gene or whether groups of genes must be considered simultaneously. Consider a gene with two alternative forms A and a that have important differential effects on the physiology of the organism. The three genotypes AA, Aa and aa will then have different probabilities of survival and reproduction expressed numerically as their *fitnesses*, conventionally denoted as W_{AA}, W_{Aa}, and W_{aa}. For a population with any proportions of the three genotypes one can calculate the average fitness in the population, W, over all individuals, by multiplying the fitness of each genotype by the genotype's relative frequency in the population and adding the results for the three genotypes. It is a basic principle of evolutionary genetics that the frequency of the different genotypes will change in such a way as to increase this average each generation and finally reach a maximum. The

Table 3.1. Hypothetical values of the fitnesses of genotypes when the genes are causally independent.

	AA	Aa	aa
BB	0.60	0.70	0.80
Bb	0.75	0.85	0.95
bb	0.80	0.90	1.00

more fit genotypes leave more offspring and are represented by a larger proportion in each succeeding generation, so the average fitness of the population increases.

Suppose there is a second gene with two alternative alleles, *B* and *b*, present in the population. Its three genotypes *BB, Bb,* and *bb* will also have different fitnesses, and the frequencies of these genotypes will also change so as to increase the average fitness of the population. But every individual in the population has one of three genotypes with respect to the *A,a* gene and one of three genotypes with respect to the *B,b* gene, so there are nine different genotypes in the population, each with its own reproductive fitness. What do the dynamics of gene evolution look like in such an instance? The answer depends upon how the fitnesses of the nine genotypes are related to each other. One possibility is that the fitness differences between the genotypes at each gene are unaffected by the other gene. An extremely simplified case is illustrated in Table 3.1. The fitness of *Aa* is exactly intermediate between the fitnesses of *AA* and *aa* irrespective of the genotype at the *B,b* locus, and, reciprocally, the fitness of the *Bb* genotype is exactly intermedi-

Table 3.2. Values of the fitnesses of nine genotypes of the chromosomal polymorphisms *bl,Bl* of the chromosome CD and *td,Td* of the chromosome EF in the Australian grasshopper *Moraba scurra.*

	Chromosome CD		
Chromosome EF	*bl/bl*	*bl/BL*	*BL/BL*
td/td	0.791	1.000	0.834
td/TD	0.670	1.006	0.901
TD/TD	0.657	0.657	1.067

ate between those of *BB* and *bb* irrespective of the situation at the *A,a* locus. In this case each gene will evolve independently of the other and the frequency of the *A* allele will increase every generation at the *A,a* locus, as will the frequency of the *B* allele at the *B,b* locus.

If, however, there is some kind of physiological interaction between the two genes, then the fitness differences at one locus may even change direction depending on the genotype at the other locus. Such a case is illustrated in Table 3.2, based on actual data on genetic polymorphisms of two different chromosomal variations, the *bl,BL* on chromosome CD, and the *td,TD* on chromosome EF, in an Australian grasshopper, *Moraba scurra.* As the table shows, the fitness differences for one of these genetic variations are greatly affected by the genotype at the other one. The most fit genotype on EF chromosome is *td/td* when the other chromosome is *bl/bl*, but the most fit genotype on the EF chromosome is *TD/TD* when the other chromosome is *BL/BL*. To analyze what will occur in an evolving population we need to consider the average fitness of the population as we vary both genetic entities simultaneously.

We can calculate the average fitness of the population for any combination of frequencies of the two different chromosomal systems and can represent that average fitness as the height above a plane whose dimensions are the frequencies of the BL and TD chromosomal types. This is shown in Figure 3.1 in a diagram that is like a geographer's topographic map. The frequencies of the *BL* and *TD* chromosomal variants are given like latitude and longitude on the two axes, and the mean fitness of each population composition is the height above the plane of the page, pictured by connecting points of equal height, as in a topographic map of a mountain range. There are two fitness "peaks" (P), one at the lower right-hand corner where both the *TD* and the *BL* chromosomes have become 100 percent in the population, and one at the upper edge where there are no *TD* chromosomes and about 55 percent *BL* chromosomes. The "valleys" of fitness (V) are in opposite corners, at lower left and upper right.

The dynamical rule is that the population changes in such a way as to climb a "peak." But which "peak"? There are two possible final outcomes of the evolutionary process, and which one will occur depends on the initial genetic composition of the population. The lines with arrows show the predicted evolution of the population from different starting conditions. Notice that a very small difference in starting conditions in the upper right-hand corner (paths 2 and 4) can result in extreme differences in final outcome. Path 3 illustrates how the climber of this adaptive mountain range can arrive at a shoulder between two peaks and become stranded there because the topography is level at this precise point. Path 5 illustrates a further compli-

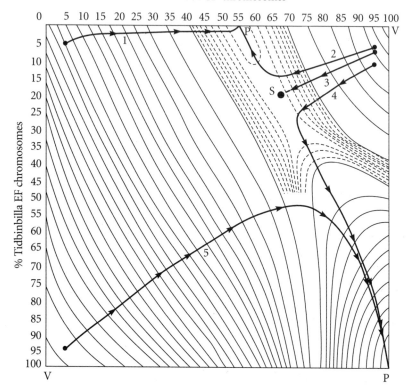

% Blundell CD chromosomes

Figure 3.1. Fitness topography for the two chromosomal systems in the grasshopper *Moraba*. The horizontal axis is the frequency of the BL form on chromosome CD, and the vertical axis is the frequency of the TD form on chromosome EF. The topographic lines connect genetic compositions of equal average fitness. "Peaks" of fitness are marked P, "valleys" are marked V, and a saddle is marked S. The arrows show the predicted paths of genetic evolution starting at different points.

cation. During the course of the evolution along this path, which is smoothly climbing the adaptive surface, always going up in average fitness, the frequency of the *TD* chromosome at first decreases from about 95 percent to about 55 percent and then reverses direction and increases to 100 percent. If we did not know of the existence of the *bl,BL* system we would say that the fitnesses of the *td,TD* system had reversed so that there must have been some change in environment. In a sense this is true, because the *bl,BL* system is a *genetic* environment for the *td,TD* pair, and as the frequencies of the genotypes at the *bl,BL* system are changing, the environment of the *td,TD* pair is constantly changing and so the fitnesses with respect to the genotypes for that polymorphism are changing. This complex behavior of the separate genetic entities *bl,BL* and *td,TD* is a consequence of our cutting up the genome of the grasshoppers in a way that is inappropriate for the problem. While the *bl,BL* and *td,TD* polymorphisms are distinct from the standpoint of identifying them genetically, they form together a single physiological unit with nine alternative forms, each with its own characteristic fitness. There is no biological reality to fitnesses calculated for the two chromosomal systems separately.

We need to be very careful about the lesson we draw from this case. It does not illustrate that parts of the genome of an organism cannot ever be treated as separate causal elements. Sometimes they can and sometimes they cannot, depending upon which genetic differences in which species living in which circumstances are being considered. There are no universal rules for cutting up organisms. In the United States, which is broken up into separate state governments with different laws, it is often impossible to say what law will hold without knowing in which state the question has arisen. When asked a question, an American lawyer will reply, "It depends on the jurisdiction." So too in biology, it depends on the jurisdiction.

∾ The alternative possible outcomes of the evolutionary process illustrated by the chromosomal variations of *Moraba* are not an isolated peculiarity, but are generally characteristic of evolutionary changes. The machine model for life has led biologists to ignore one of the common characteristics of many physical systems, their dependence on initial conditions. If I say that I traveled eighty miles west and twenty miles north, it is impossible to determine that I arrived in Brattleboro without knowing that I started in Boston. This dependence on initial conditions is not a characteristic, however, of the typical machine, whose operation is generally independent of its individual development or the history of its invention.[1] The mechanic does not need to know how a car assembly plant works or the history of the invention and development of the internal combustion engine in order to know why a car has a fuel pump. But the biologist is not a mechanic. It is impossible to understand the situation of living organisms without taking into account their history.

All species that exist are the result of a unique historical process from the origins of life, a process that might have taken many paths other than the one it actually took. Evolution is not an unfolding but an historically contingent wandering pathway through the space of possibilities. Part of the historical contingency arises because the physical conditions in which life has evolved also have a contingent history, but much of the uncertainty of evolution arises from the existence of multiple possible pathways even when external conditions are fixed. It is a prejudice of evolutionists who give adaptive explanations of the features of organisms that every difference between species must be a consequence of different selective forces that operated on them. However, in the simple case of the chromosomal variations in *Moraba* that is not true. The fitness differences among the nine genotypes are fixed constants and do not differ between a population following path 2 and one following path 4. Yet the outcomes are radically different. Populations subject to identical selective conditions may arrive at quite different evolutionary endpoints, so that the observation that two species differ is not *prima facie* evidence that they were adaptively differentiated. There are many cases in which related groups of species have a great variety of forms of the same basic feature, but in which there seems no way to provide a special story of selection for each form.

For example, the ceratopsian dinosaurs had bony horns on their heads, much like modern rhinoceroses, and extensions of the rear of their skulls to make a large bony frilled collar (Figure 3.2). It is reasonable to speculate that these skull ornaments of the herbivorous ceratopsians functioned as protective devices against the attacks of carnivorous predators or in aggressive confrontations between individual ceratopsians themselves. What is not so easy to explain is the immense variation in the number and size of the horns and the extent of the bony collar from one species to another. There was no general temporal trend to increase the size of these features, and small-horned ceratopsians lived at the same time as large-horned species. The modern equivalent is the presence of two-horned rhinoceros in Africa and one-horned rhinoceros in India. Do we really want to argue that something in the environment of Africa favors two long slender nose horns while the selective

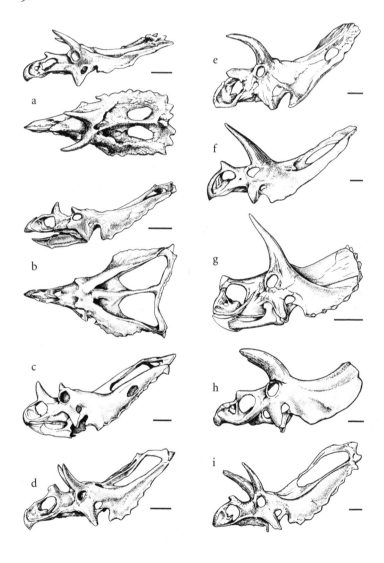

conditions in India favor one short fat horn? The simplest explanation is that these are two alternative outcomes of the same selective process beginning with somewhat different initial genetic conditions, as in the grasshopper case.

The uncertainty of selective outcome arises from an even more elementary process, the origin of variation itself. The Darwinian mechanism for evolution involves the selection of existing genetic variation, increasing the frequency of some genotypes and reducing the frequency of others. Selection cannot occur for a particular characteristic if some genetic variation in the direction of that characteristic is not present in the population. It is useless to argue that natural selection would favor a vertebrate with two wings in addition to its four limbs, because no such variation in the genes controlling early segmentation has occurred, or if it has it has not been of a kind that would allow regular development to proceed.

Genetic variation depends on the process of mutation, and mutations are rare events. Any particular new DNA mutation will occur only once in about 100 million gametes. Moreover, when a single mutation occurs in a single newborn, even if it is a favorable mutation, there is a fair probability that it will not be represented in the next generation because its single carrier may not, by chance, pass it on to its few offspring. The time between the origin of a species and the time that a mutation of just the right sort occurs and reaches a high enough frequency

Figure 3.2. Skulls of a variety of ceratopsian dinosaurs showing different sizes of horns and collars.
From *The Dinosauria* by Weishampel et al., © 1990 by the Regents of the University of California and the California University Press. Used with permission.

to be significant in the selective process is of the same order as the total lifetime of the species, around ten million years. So most mutations that would have been selected if they had occurred are never seen. A species must make do with the variation it actually has.

Moreover, the mutations that can occur in a species are conditioned by its current genetic state. Each mutation is a single substitution in DNA. To produce a selectively useful genetic variant from the current DNA sequence may require not one mutation but several, each of exactly the correct sort. Given that vertebrates are four-limbed animals, it might take many mutations, no one of them useful in itself, to reach a genetic variant that could be the basis for adding wings while keeping legs. In Shakespeare's *Henry IV, Part One* the pompous Owen Glendower boasts of his powers: "I can call monsters from the vasty deep." To which Hotspur replies, "Why so can I, and so can any man, but will they come when you do call for them?" Selection may call, but there may be no mutations to answer.

➣ A consequence of the intermediate size and internal heterogeneity of living organisms is that they are the nexus of a very large number of weakly determining forces. The elliptical motion of the planets is determined by their masses and by their velocities and distances from each other and from the sun. At the other end of the size scale, the chemical and physical properties of atoms are consequences of the number of electrons and nucleons of which the atoms are composed. The properties and motions of very large and very small effectively homogeneous systems are determined by a small number of interacting forces, any one of which has a major effect on the system when it is perturbed. The study of such systems has been the model for physical science, and the immense success of physics and chemistry in predicting or manipulating the world is a consequence of the major causal effect of single factors. By reproducing fixed values of a few manipulable variables scientists can effectively replicate the behavior of the systems. There may sometimes be great sensitivity of the final state to small differences in initial values, but this is a question of accuracy rather than complexity of causation. So, a mid-course correction may be necessary in sending a space probe to Saturn, because of the immense distance to be traveled, but the forces of gravitation, momentum, and inertia are sufficient to make the pathway of the probe easily predictable and manipulable.

The problem for biology is that the model of physics, held up as the paradigm for science, is not applicable because the analogues of mass, velocity, and distance do not exist for organisms. Organisms are of intermediate size and take odd shapes. As a result it is not the first book of Newton's *Principia*, which deals with idealized systems in vacuums, but the second, which discusses friction, buoyancy, and the movement of real objects in real media, that is most relevant to them. Organisms move in a viscous medium; they suffer friction; they are too small and too distant from each other to interact gravitationally; their collisions are not elastic; their shapes, masses, and centers of gravity are changing; if they live in water they are buoyant; their paths are constantly being influenced by external and internal forces. The characteristic of a living object is that it reacts to external stimuli rather than being passively propelled by them. An organism's life consists of constant mid-course corrections.

Organisms are also extremely internally heterogeneous.

Their states and motions are consequences of many intersecting causal pathways, and it is unusual that normal variation in any one of these pathways has a strong effect on the outcome. To be ill is precisely to be dominated by a single causal chain. To be obsessed by an *idée fixe* which motivates all one's actions, or to be convinced that all behavior on the part of others, without distinction, is hostile, is a form of mental illness. To be a victim of a malfunctioning liver or kidney or a growing tumor, or even to suffer from a non-life-threatening respiratory infection, is to be dominated by a single abnormal physiological element. Indeed, we may define "normality" as the condition in which no single causal pathway controls the organism.

The multiplicity of causal chains, all of weak individual influence in their normal condition, presents a special difficulty for the attempt to understand life processes. All attempts to understand causes must necessarily involve the observation of variations. It is not possible to ascribe a cause to some effect unless the putative cause and its effect can be seen to vary together. The standard method of analysis in genetics, for example, is to use the developmental and physiological variations caused by gene mutations to assign causal roles to the genes. There are two alternative pathways open for the causal study of variation. One is to observe systems in their natural state and to observe the correlations between various aspects of their condition. This is, in fact, the method employed by the founders of modern celestial mechanics, Galileo, Kepler, and Newton. Kepler's Laws are generalizations about the law-like behavior of planets derived from the differences in the orbits of different planets, and this remains the chief technique employed by modern cosmologists, who must, after all, take the universe as they find it.

For biologists, this is the comparative method of natural history. It is the source of famous "laws" or "rules" of variation, which are really only expressions of tendencies rather than rigid relationships, like Bergmann's Rule that closely related warm-blooded animals are larger in colder regions than in warmer regions. The causal explanation offered is that the ratio of surface area to volume decreases with increasing size and that the preservation of body heat is a problem for animals in very cold places, so the smaller the relative surface area the better. Large animals usually have longer lifetimes than smaller ones because they have lower metabolic rates, but not always. Species tend to be more variable in the geographical region near their center of distribution, but there are many exceptions. The number of rules of this kind in biology, established on a purely correlational basis from observed natural variation, is not very great, and when general tendencies are observed, the causal explanations are not easily testable precisely because all the available data are already absorbed in the making of the generalization.

The failure to find Keplerian regularities from natural observations is one consequence of the multiple causal pathways. Causal claims are usually *ceteris paribus,* but in biology all other things are almost never equal. The natural differences in effects observed among organisms do not usually have sufficient regularity with respect to the natural variation of individual causes, because these individual causally relevant variables are each too weak in their effects to dominate the large number of other variables. As a result, biologists, like other scientists, resort to experiments in which deliberate perturbations are introduced. In such experiments, however, unlike the situation in the study of simple physical systems, there is a serious scaling effect. If a normal organism is one that is the nexus of many weakly determining forces, then the experimental organism which is perturbed strongly enough by a single causal deviation to show a reliable effect is an abnormal organism. The problem for biology in each case is whether the abnormally large experimental perturbations really reveal the causes of the smaller natural differences.

This problem is acute in genetics. A drastic mutation of one of the homeobox genes in *Drosophila* certainly reveals that the reading of this gene plays a central role in the development of wings in the insect. But it does not explain the normal variation in wing size unless it can be also be shown that such variation is in fact associated with minor differences in DNA sequence in this gene, and that there are not other important sources of variation. Although there is now evidence that variations in DNA sequence of homeobox genes do, in fact, account for some of the natural variation in wing size, it would be surprising if they were found to account for most or all of that variation in different populations of *Drosophila*.

At one time an important experimental technique for exploring the genetic causes of normal morphological variation was the selection experiment. In this method two breeding lines of organisms are established, one in which the individuals with the largest measurement, say the longest wings, are the parents of each successive generation, and one in which the individuals with the smallest wings are the parents. After a sufficient number of generations the "high" line and the "low" line will be quite different in wing size. These are then used in crosses with a variety of special genetic stocks to determine where on the chromosomes the genes lie that differentiate to the two lines. These are the "genes for wing size." A common feature of such experiments was that when they were repeated in independent experiments the "genes for wing size" turned up in different chromosomal locations in each experiment.

The modern equivalent of these experiments is the attempt to locate the genes for mental conditions like schizophrenia or bipolar syndrome by observing the way in which these phenotypes pass down in pedigrees together with known chromosomal markers. The results are equally inconsistent. In one large family pedigree "the gene for bipolar syndrome" will be definitively located on one chromosome, while in another family it will turn up on a different chromosome. Putting aside experimental and statistical problems with the observations, this apparent inconsistency is entirely reasonable given the imprecision of the definition of the trait and the multiple genetic pathways that must contribute to the formation of the central nervous system. We should not expect that single drastic genetic changes produced either experimentally or by the bad luck of naturally occurring mutations will account in specific cases for most, or even any, of the normal variation we see in nature.

What is true for genetic perturbations is also true for chemical or temperature or any other external changes. In their normal state organisms are buffered against the effects of many internal and external changes by homeostatic regulatory devices. The control of body temperature in mammals by changes in heart rate, blood sugar level, hormone levels, dilation and contraction of blood vessels and skin surface muscles, and the position of body hairs is a well-known example. Another is the large number of feedback devices within cells that keep their division times and metabolic rates within narrow limits and that may produce a constancy of developmental pattern despite the existence of genetic and environmental factors tending to perturb them. But all of these homeostatic devices work only within certain limits of perturbation, and if the disturbance is too great the organism will show a response. The control of body temperature is lost after relatively brief immersion in freezing or extremely hot water. The same limitation exists for developmental regulation. All normal *Drosophila* have four large hairs, the *scutellar* bristles, on their backs. If a drastic mutation is introduced into the flies' genome, the number of scutellar bristles is reduced, but considerable variation appears among individuals, some having no bristles, some one, and some two. Selection experiments have shown that this variation is largely a consequence of genetic differences that were already present among the normal flies, but that could not be observed until the developmental system was severely stressed by the addition of the drastic mutation. Provided that development is not pushed beyond its usual pathway by other perturbing forces, there is no developmental effect of this variation in genes.

The existence of homeostatic devices of limited range means that biological systems have thresholds for the effect of causal variations. Small natural variations along causal pathways will be without effect, while extreme experimental perturbations put the organism in a range of conditions unrelated to its normal functioning. Evolutionary geneticists would like very much to know whether natural selection discriminates among certain observed genetic variations in species. If so, it is almost certain that the changes caused in the composition of populations are too slow to be observed in an experimenter's lifetime. The physiological differences between the genotypes can be greatly exaggerated experimentally by the appropriate external conditions, but what will that tell us about the causes of a population's genetic composition in nature?

The limitation of experimental biology to manipulating one or a small number of causes by large perturbations has had a profound effect on the kinds of explanations that are offered by biologists. The methodological limitations of experiments are confused with the correct explanations of the phenomena. The constant reiteration of the claim that genes determine organisms is a consequence of the ease with which major genetic changes can be induced in experiments and the large effects that those changes have on the experimental objects. Moreover, only those phenomena are considered that lend themselves to the method. Developmental geneticists ask questions about the differentiation of anterior and posterior ends of animals and the formation of major body segments in between because single major gene defects can be found that alter that process. They do not know how to ask why different individuals have heads and legs of different sizes and shapes, even individuals of different species. So they never ask.

≫ Yet another failure of the classical machine model in biology arises from some difficulties of separating causes from effects. The concept of feedback loops has now firmly entered the treatment of physical systems as a result of the development of cybernetic and control theory. We are used to the idea that a perturbation in one part of a connected system may be the cause of an effect in another part, which then becomes a cause for a change in the first part. This mode of explanation is the rule in much of cellular physiology and metabolism, in some models of the genetic control of early development, and in aspects of neurobiology and gross physiology. The domain of such explanation in biology is just that domain in which the old simple machine metaphor was most directly applicable. The cell or the body is no longer seen as a collection of gears and levers but as a system of signaling pathways that allow for feedback to maintain stable states or rates and timing of flows. Always there is a model. It used to be a clock; now it is a servo-motor.

"Regulation" is one of the most common words in functional biology. In other branches of biology, particularly ecology and evolution, however, there is no simple electro-mechanical picture to provide the model for the system. As a result the abstract features of the older machine metaphor continue to dominate the understanding of causes and effects. In Chapter II I discussed the model of the world in which there is an outside force, the preexistent environment, that dictates the "problems" organisms must solve, and inside forces of variation that generate the organisms' "solutions" to the "problems." Organisms map the autonomous external changes in the world. The external environment in such a view is the cause, the evolved morphology, physiology, and behavior of the organism is the effect, and natural selection is the mechanism by which the autonomous external cause is translated into the effect. But, as shown in Chapter II, this asymmetrical cause-and-effect picture does not capture the truth about the relation between organisms and their environments. Just as immediate changes in organisms are the effects of natural selection in a given immediate environment, those changes become the causes of changes in that environment. I argued in Chapter I that an organism is not coded in its genes because the environment in which development occurs must be taken into account. But the argument of Chapter II suggests that, paradoxically, the *environment* is coded in the organisms' genes, since the activities of the organism construct the environment.

Taken together, the relations of genes, organisms, and environments are reciprocal relations in which all three elements are both causes and effects. Genes and environment are both causes of organisms, which are, in turn, causes of environments, so that genes become causes of environments as mediated by the organisms.

The classical picture of evolution can be represented formally as a pair of differential equations in time:

(1) $dE/dt = f(E)$
(2) $dO/dt = g(O,E)$

Equation (1) states that there is some change in environment over time that is entirely a function of environmental variables, and equation (2) states that the change in the organism is a function both of the current state of the organism and of environment. Because equation (1) is a function only of environment, it can be solved to provide a complete temporal history of the environment. This solution, when substituted into equation (2), will then give a complete evolutionary history of the organism, driven entirely by the autonomous environmental process. The actual situation, however, is that evolution is a pair of *coupled* differential equations:

(3) $dE/dt = f(O,E)$
(4) $dO/dt = g(O,E)$

so that the histories of both environment and organism are functions of both environment and organism. The equations must be solved together as a coupled pair describing the *co-evolution* of organism and environment in which both are acting as both causes and effects.

➣ There is a final question about causes that arises both from the multiple causal pathways in biological systems and from their material nature as functionally coordinated systems. This is the distinction between *causes* and *agencies*. The confusion between causes and agencies is nowhere more apparent than in medical science. It is common to speak of the causes of death, which, in industrial societies, are chiefly heart disease, stroke, and cancer. Immense effort is put into finding the mechanisms of these diseases in the hope that they can be prevented or so ameliorated that people will not die of them. But suppose that all forms of cardiovascular disease and cancer could be successfully treated or prevented. Does that mean that we would not die?

In ordinary causal analysis, we distinguish between necessary and sufficient causes. If something is a necessary cause of an effect, then if we avoid the cause we avoid the effect. We can eliminate a sufficient cause, in contrast, without preventing the effect, because some other cause may take its place, but if the sufficient cause is present, the effect will inevitably follow from it. In this simple analysis, however, cardiovascular disease and cancer are neither necessary nor sufficient causes of death. Having either of these diseases does not guarantee that one will die of them, nor does freedom from these diseases provide immortality. They are neither necessary nor sufficient causes of death. They are two out of a large set of alternative proximate causes of death. It is possible to avoid any one of these alternatives, but it is not possible to avoid them all. If one does not die of one cause one will die of another.

But why should this be the case? If these causes of death are functionally independent, then it ought to be possible to escape them all, and, indeed, the claims made by medicine imply this possibility without explicitly stating it. Medical scientists speak of "preventing" deaths by curing disease, but the evidence is that death cannot be prevented, only postponed at best. Moreover, the postponement has not been as effective as is sometimes claimed during the last fifty years of great progress in physiology, cell biology, and medicine. The expectation of life at birth for a white male in the United States has increased by seven years since 1947, but this is not because people are living to a great age. Life span has not increased, and the number of years of further life expected for the author of this book, a seventy-year-old white male, has only increased by about two years. It must be that although the proximate causes of death can be dealt with, death itself cannot.

So there must be a cause of death as a phenomenon, as distinct from the individual cases, which are better thought of as "agencies." Agencies are alternative paths of mediation of some basic cause, a cause that always operates, although through different pathways. If the cause does not operate through one agency it must operate through another. In this light the cause of death is that living organisms are electro-mechanical devices, made up of articulated physical parts which, for purely thermodynamic reasons, eventually wear out and fail to function. Different parts wear out at different times in different individuals, and some parts are more prone to failure than others, or are located in the functional articulation at a place that is more critical. My motorcar must eventually go to the scrap heap because either the engine or the transmission or the electrical system will fail from wear and decay. I could, of course, keep it forever by replacing each part over and over again, but it is not clear that after every part had been replaced it would be the same motorcar. There is a story in rural Vermont of a man who claimed to have had an axe in his family for 150 years. When asked how that was possible, he said that his family had taken very good care of it by providing seven new handles and three new heads.

The distinction between causes and agencies can have important effects on the actions that are taken to intervene in human affairs. In the nineteenth century in Europe the chief "causes" of mortality were not cardiovascular disease or cancer, but infectious diseases. The mortality statistics show that the most important killers were diphtheria, smallpox, tuberculosis, bronchitis, pneumonia, and, in children, measles. At the time of the first systematic recording of these sources of mortality in the 1830s, the death rates from all of these diseases were decreasing, and 90 percent of the decrease had already occurred by the time of the First World War. What was the reason for this dramatic change? It was not the discovery of the pathogens, because there was no observable effect on these death rates after the germ theory of disease was announced by Robert Koch in 1876. It was not the introduction of modern drug treatments, because from 90 to 95 percent of the reduction in death rates from these "causes" had already occurred when antibiotics were introduced after the Second World War. It was not improvements in sanitation, since all these principal killers were airborne, not waterborne, diseases. Nor could the change have been entirely caused by measures designed to prevent diseases from spreading. Measles was the principal fatal disease of children in the nineteenth century, but when I was a child no one died of measles, although every child contracted it.

The most plausible explanation we have is that during the nineteenth century there was a general trend of increase in the real wage, an increase in the state of nutrition of European populations, and a decrease in the number of hours worked. As people were better nourished and better clothed and had more rest time to recover from taxing labor, their bodies, being in a less stressed physiological state, were better able to recover from the further severe stress of infection. So, although they may still have fallen sick, they survived. Infectious diseases were not the causes of death, but only the agencies. The causes of death in Europe in earlier times were what they still are in the Third World: overwork and undernourishment. The conclusion to be drawn from this account is that the level of mortality in Africa does not depend chiefly on the state of medicine but on the state of international production and exchange, although it would be absurd to say that medical care is irrelevant.

The same distinction between causes and agencies is relevant to problems of pollution and the management of waste. When popular and legal action is successful in preventing a particular industrial process that poisons workers or destroys resources or accumulates non-degradable wastes, industry switches to a different process in which other poisons or wastes are produced and other resources consumed. Paper consumes trees and puts sulfites into the water and air. Its replacement by plastic consumes petroleum and creates a non-degradable end product. Miners no longer die of black lung from coal mines as coal is replaced by petroleum. Instead they die of cancer induced by the products of refineries. Sulfites, deforested mountainsides, non-degradable waste dumps are not the causes of the degradation of the conditions for human life, they are only its agencies. The cause is the narrow rationality of an anarchic scheme of production that was developed by industrial capitalism and adopted by industrial socialism. In this, as in all else, the confusion between agencies and causes prevents a realistic confrontation with the conditions of human life.

IV DIRECTIONS IN THE STUDY OF BIOLOGY

⇒ The earlier chapters in this book have a distinctly negative flavor. They are devoted to explanations of the way in which a reductionist approach to the study of living organisms can lead us to formulate incomplete answers to questions about biology or to miss the essential features of biological processes or to ask the wrong questions in the first place. It is easy to be a critic. All one needs to do is to think very hard about any complex aspect of the world and it quickly becomes apparent why this or that approach to its study is defective in some way. It is rather more difficult to suggest how we can, in practice, do better. It is useless to call in general terms for some more synthetic approach or to say that somehow we need a new insight.

One alternative for dealing with the complexities of living systems is to go to the other pole from unremitting reductionism and to claim that the world is a single, unanalyzable structure of interactions which cannot be broken down into parts in any way without destroying what is essential to it. There is some element of this holism in parts of the ecology movement, reaching its most extreme expression in the "Gaia" hypothesis, according to which the biosphere, atmosphere, and geosphere form "a totality constituting a feedback or cybernetic system which seeks an optimal physical and chemical environment for life on this planet."[1] But extreme holism fails as a program for biology for two reasons. First, even if it were true that everything is strongly connected to everything, that should not be confused with the methodological claim that no success at all in understanding the world or in manipulating it is possible if we cut it up in any way. Such a strong methodological claim we know to be wrong as a matter of historical experience. Whatever the faults of reductionism, we have accomplished a great deal by employing reduction as a methodological strategy. Second, the holist claim is wrong as a description of the world. Everything is not effectively connected to everything. While gravitational perturbations do indeed spread out into the indefinite distance, one can stir a flower without troubling a star, because gravitation is a weak force that decreases as the square of the distance between objects. The world is divided into nearly independent subsystems within which there are effective interactions but between which there are no palpable relations.

The problem of science is to find the boundaries of those subsystems. It is not the case that the extinction of a single species will have effects that propagate throughout the entire living world, but the disappearance of that species would certainly have a major effect on some other species that depended upon it for food. Moreover, the boundaries of the subsystems within which there are significant interactions change with circumstances. The loss of the tip of my left little finger would not change my life significantly, but it would if I were a concert pianist. The invariable presence of the little finger among primates, despite the great variation in digit pattern among terrestrial vertebrates, certainly points to the inclusion of that digit in the effective grasping unit that was important at an earlier stage of human evolution. The delineation of effective subsystem boundaries is a major practical task for the biologist in all circumstances. Obscurantist holism is both fruitless and wrong as a description of the world.

Another alternative is to search for general systems of explanation of which the diversity of biological phenomena will appear as special cases. Over the last thirty-five years there have been three attempts to bring biological phenomena under the aegis of very general properties of systems that are changing in time. They are the Three C's: catastrophe theory, chaos theory, and complexity theory. All are attempts to show that extremely simple relationships in dynamical systems will lead to what, at first sight, seem unpredictable changes and extraordinary diversity of outcome.

Catastrophe theory, developed by the mathematician René Thom in the 1960s, shows that in some systems that are changing in time according to quite simple mathematical laws, the changes observed may be continuous and gradual deformations of the state at a previous instant, but that at a critical point the entire shape of the system will undergo a "catastrophic" change and then continue its development along a totally new pathway. The classical example is the breaking wave, in which a swell develops into a deep convex curve by a continuous deformation of shape whose tubularity is suddenly lost at a critical point when it comes crashing down. An obvious biological analogue is the complex change in shape that takes place during the development of an embryo. Perhaps the foldings, slidings, separations, and fusions of tissues can be discovered to be the continuous unfolding of a simple law of change. The biological practitioners of catastrophe theory hoped that it would provide the explanation of, among other things, the extinction of species. Fewer than one one-hundredth of one percent of all the species that have ever lived are still extant, and all species eventually become extinct. A hope of catastrophe theory was to show that extinction was simply the consequence of the same demographic and ecological laws that allowed a species to spread in the first place, but there is currently no trace of this theory in biological practice, and truly "catastrophic" external events, meteor impacts, have probably been responsible for a major part of species extinctions.

In the 1980s chaos theory was developed in meteorology to show that some very simple dynamic systems may go to equilibrium or undergo regular oscillations in one range of parameters, but in other ranges will pass from one state to another in what appears to be a totally random fashion. These apparently random changes of state, however, can be perfectly predicted from the simple equations of motion of the system. So an uncertain and diverse world is really the deterministic solution to a trivially simple equation, and if only we knew the equation and the values of the parameters we could predict the entire history of the system. Ecologists have built simple models of population growth that exhibit chaotic behavior in time as a way of explaining the seemingly random changes that occur in the abundance of species. So, the occasional unpredicted outbreaks of huge populations of the gypsy moth in New England would not require any explanation involving special conditions of the environment or of the abundance of other species. They are the seemingly chaotic outcome of a simple deterministic process. Outside of pure speculation and the ad hoc adjustment of the parameter ranges of simple population models, however, chaos theory has not helped us to understand what is really going on in nature, because we do not know how to discover the "laws" of population growth, if they exist. By postulating simplified models of population growth that predict chaotic regimes we cut ourselves off from any program of further investigation of other causes of temporal variation in population numbers. Do gypsy moths have occasional unexplained outbreaks? Well, what did you expect? After all, they are subject to chaotic population growth.

The most recent entry into the struggle to understand biological phenomena is the hope that complex systems have special laws that originate in the multiplicity of interactions among many parts, laws of complexity itself.[2] These laws of complex wholes would arise, not from any new forms of interaction between elements of living systems as distinct from inorganic physical relations, but from the sheer number of elementary interacting parts. Thus, if there are many genes relevant to the development of some feature of an organism, and if the transcription of these genes is connected by a network of multiple pathways of simple "off-on" signals, some sort of laws of development will emerge. It remains to be seen whether this approach to complexity will illumine the problems of biology. So far, we have only a speculative enterprise.

Rather than searching for radically different ways of studying organisms or for new laws of nature that will be manifest in living beings, what biology needs to do to fulfill its program of understanding and manipulation is to take seriously what we already know to be true. It is not new principles that we need but a willingness to accept the consequences of the fact that biological systems occupy a different region of the space of physical relations than do simpler physico-chemical systems, a region in which the objects are characterized, first, by a very great internal physical and chemical heterogeneity and, second, by a dynamic exchange between processes internal to the objects and the world outside of them. That is, organisms are internally heterogeneous open systems.

➣ One consequence of the internal heterogeneity is that function cannot be understood without information about shape and form. It is a requirement of the study of life that the shapes and spatial relations of populations, individuals, cells, and molecules be taken into account in the study of function. Biology began as a study of the shapes of plants and animals, and the functions of parts of organisms were understood as intimately related to their shapes. The nineteenth-century sciences of phrenology and criminal anthropology were based in the belief that character and cognitive function would leave their mark in the shapes of skulls and noses. Evolutionary biology and the systematic classification of organisms were, until recently, based primarily on shape, in large part because that was the only characteristic of organisms that was preserved in the fossil record or in collections. But, as chemistry invaded biology toward the end of the nineteenth century, with the consequence that organisms came to be seen as collections of molecules, questions of form were de-emphasized in favor of the study of molecular reactions. Form again came to play some role in reductionist biological explanation with the development of the biochemistry of macromolecules in the last hundred years. Clearly the shapes of enzymes and their substrates are critical to their chemical interactions, and we cannot understand muscle contraction or the function of membranes or the transport of oxygen by hemoglobin without reference to how the relevant molecules fill space and are juxtaposed in relation to each other.

The culmination of that interest in form came with the discovery of the three-dimensional structure of DNA and the consequences of that structure for explanations of DNA replication and coding. Yet, ironically, the understanding of the role of DNA in biology has led to a scheme of explanation that pays minimal attention to molecular structure and the spatial relations of molecules. The central dogma of molecular biology is that the chemical sequence of nucleotides in DNA determines the chemical sequence of amino acids in proteins which in turn uniquely determines the function of the protein. The new central dogma of developmental genetics is that the development of the shape of an organism can be understood by the complete set of signaling pathways among genes, which gene "turns on" which other genes. Of course, these explanations do not ignore shape entirely. We know that the blocking or promotion of transcription of a gene depends upon the coiling of DNA on chromosomes and how DNA-binding proteins can fit into the major groove of the double helix. It is part of the substrate of our understanding of biochemistry that the folded three-dimensional shape of a protein is critical in its function. Yet that understanding does not enter in an integral way into biological explanation.

The full explanation of the path between gene and organism needs to include known phenomena that influence the way in which the string of amino acids coded by the gene becomes a protein, that is, a folded three-dimensional structure. The sequence of amino acids is insufficient to explain this folding, and there are many alternative stable folded states for any sequence, only one of which is the physiologically active protein. The particular folded state at which the cell arrives during its manufacture of proteins depends partly on the exact DNA sequence that codes for the amino acid sequence. The DNA code is redundant. That is, there are different DNA codes that correspond to the same amino acid. The different codes for the same amino acid influence protein production. During the process of gene transcription the DNA code is copied into an RNA, messenger RNA, which then enters into the process of protein production. The exact nucleotide sequence of this message influences the speed with which different bits of the sequence get translated into sections of amino acid sequence during the construction of the protein, and those variations in speed influence the folding of the protein as it is produced. So, if we want to understand which protein is produced from a gene, we must investigate the complete chain of production in its spatial and temporal detail.

The folding of proteins also depends upon the cell environment. Human insulin for pharmaceutical purposes is now produced in fermentation vats by bacteria that carry the human insulin gene. When this gene was first transferred to bacteria, the protein that was produced did not have physiological activity even though it had the correct amino-acid sequence. It turned out that the protein was folded incorrectly by the bacterial cell, a problem that was solved by changing the culture conditions. This is a model for one way in which the environment may interact with the genotype in the organism's development.

There have been a number of striking demonstrations of how three-dimensional molecular structure can illumine biological explanations, for example the complete elucidation of the structure of the molecular motor that drives the cilia of bacteria in their circular motion, or the demonstration of how the three-dimensional form of the protein of influenza viruses has evolved to escape from antibodies produced by their host.

Recently a case has been discovered in which a single amino acid substitution in an enzyme changes completely the kind of chemical substrate on which the enzyme operates and the kind of chemical bond that it breaks. That change is explained by very small changes in the angles between other amino acids, which now bind a water molecule that participates in the new enzymatic reaction. This change confers resistance to insecticide on the enzyme's carrier, the blowfly.[3]

These understandings, however, have not penetrated into the main structure of biological explanation. It is not possible to read the literature of, say, developmental genetics without noticing that such observations on structural change are peripheral to the main body of investigation of the relation between gene, protein, and organism, reserved as the concern of a small number of specialists in molecular structure. What is needed is to move the issue of structure from the peripheral realm of a few special cases to a central concern of investigation at the molecular level.

Similarly, the shape and internal arrangement of cells must become a central feature of the explanations of development. An immense amount is known about the internal structure of cells and about the localizations of various molecules within them and on their membranes. Beyond the descriptive level, however, the processes of differentiation of an unspecialized cell into a mature specialized form are not understood. Yet cell differentiation lies at the basis of all development. The timing of cell division, the plane in which that division occurs, and the migration and adhesion of the resulting cells are all processes that are basic to development. The problems of cell differentiation, division, and movement cannot be solved without information about the spatial distribution and organization of molecules within cells and about how the state of a cell is influenced by neighboring cells and the surrounding environment. We need to return to the old problem of "positional information." It is all very well to say that certain genes come to be transcribed in certain cells under the influence of the transcription of certain other genes, but the real question of the generation of form is how the cell "knows" where it is in the embryo. For this purpose we do not need a revolutionary insight into the laws of biology, but only a lot of hard work.

The second consequence of the internal heterogeneity of organisms is that it is very dangerous to extrapolate from a few convenient examples to the whole of biology. As discussed in Chapter III, a result of the heterogeneity is that the normal organism is the nexus of a large number of interacting forces that are individually weak. In order to study a system of weak and interacting forces it is convenient to hold as many as possible of the causal pathways constant and to make a strong perturbation of one of them to determine its influence. This is the nature of experimental as opposed to observational science. The problem that arises is whether the effect of an abnormally strong perturbation will scale down to the normal effect of the normally weak variation in a cause on the background of all the other varying causal pathways. In simple physical systems the rules of scaling are understood, but even in complex inorganic situations like meteorology or aerodynamics, simple by the standard of biology, extrapolations are not easy. Many of the problems of understanding in biology appear when experimental perturbations, chosen because they can be studied conveniently, are extrapolated to normal circumstances.

Geneticists who use *Drosophila* as an experimental organism have at their disposal thousands of gene mutations that have been characterized and mapped on chromosomes. These mutations are described in a large catalogue where each is given a rank from 1 to 3. Rank 1 mutations are those in which the abnormal change in observable phenotype of the organism is clear and unambiguous, in which all individuals that carry the mutation show the abnormal phenotype, and in which the abnormality is apparent over a wide range of conditions of laboratory culture, or at least under an easily reproducible condition. Rank 2 and Rank 3 mutations, in contrast, appear in only some fraction of those individuals that carry the mutant genotype, and vary widely in the intensity of their expression from individual to individual even in controlled environmental conditions so that it is not always easy to decide which individuals carry the changed gene. Experimentalists use Rank 1 mutants in their experiments whenever possible, for obvious reasons of experimental ease. As a result of this practice, they come to see the world of genetic changes as if they were all Rank 1 changes, but, in fact, the vast majority of mutations in the catalogue are of lower rank. Most known mutations are not so different from the normal, do not appear at all in many of the individuals who carry them, and are quite sensitive in their appearance to temperature, age, humidity, and other non-genetic factors, and to the state of other genes, so-called modifiers.

Sometimes, because of interest in a particular gene, experimenters combine mutations of partial effect in one individual and select for enhancing modifiers to exaggerate the developmental effect. An illustration often used in textbooks shows a fly with two complete pairs of wings instead of the normal condition in which there is one pair of wings and one small knob-shaped flight balancer. Although this is commonly presented as an example of a simple mutational effect, the strain of flies with two perfect sets of wings has had three different Rank 3 mutations combined and the expression of the mutations has been enhanced by selection of the right modifier genes.

Moreover, much of the phenotypic variation that appears among individuals in a population of *Drosophila* or any other organism, variation in size, shape, physiology, and behavior, cannot be traced to any well-defined variation for a particular gene, if they are influenced by genes at all. Human skin color, for example, is clearly heritable, at least if we consider the differences between geographically defined populations, but the genes for human skin color have never been located. In contrast, the repeated natural experiment of human relocation and cultural absorption shows that the world variation in the phonemic structure of human languages is not influenced at all by genetic differences. A Finnish child adopted at birth by the Xhosa would experience no difficulty at all in making palatal clicks.

Thus the Rank 1 mutants used by the experimentalist cannot be taken as a model for most organic variation within a species. If we want to understand the causes of the differences in shape, size, color, behavior, and physiology among individuals we must be prepared to work with genetic differences at many gene loci, each of small effect, and with interactions between gene and environment like those discussed in Chapter I. There is no problem in principle in studying such quantitative gene effects, only practical difficulties. We need to determine the norms of reaction and the role of developmental noise (see Chapter I) for each case of interest. A yet deeper analysis of the interaction between genes and environment requires an understanding of the development of phenotype at the cellular and molecular level. There is no easier pathway.

A related aspect of the use of conveniently large perturbations is the use of "knockouts" in the study of development and function. Whether by means of mutations of large effect, or by chemical, electrical, or surgical manipulation, or by the study of accidental traumatic damage to the physiology and anatomy of organisms, we attempt to understand normal causal pathways by examining the effects of major disturbances. The correspondence between cognitive function and brain anatomy is studied in split-brain experiments, or in lobotomies, or in people who have suffered severe destruction of parts of the brain. The acquisition of bird song is studied by deafening birds to see how much difference aural experience makes. All studies of the genetic mediation of development are carried out using gene mutations or methods of genetic engineering that block signaling between parts of the genome or prevent the production of particular gene transcripts.

The only way to find out whether such knockout experiments have identified the elements of normal function is to analyze the small variations that are normally encountered to see whether the same causal pathways are involved. This is by no means impossible. For example, recent experiments have shown that the degree of sensitivity of *Drosophila* to environmentally induced variations in the morphology of the thorax involve, at least in part, genetic variations of a gene, *Ultrabithorax,* whose major knockout mutations make large differences in the development of that part of the body.[4] What is required is not a major theoretical insight into development, but detailed work on small effects.

≫ At first sight it may seem that there is no hope of making any constructive generalizations about how to take an organism apart without destroying the very relations we are trying to understand. Can we say nothing more than, as lawyers insist, "it all depends on the jurisdiction"? In fact, we can do better than that, because we can point to one domain of action in which it is virtually certain that small differences have significant effects on the whole organism: the domain of macromolecular structure. During the last fifteen years, molecular population geneticists have studied the variation in DNA sequence from individual to individual within species for a number of genes. Typically, there is a great deal of DNA variation: in *Drosophila* about 6 percent of the nucleotide positions in a gene will show some variation among individuals even in quite small samples. However, despite the fact that roughly three-quarters of all random mutations in DNA should result in an amino acid substitution, virtually none of the observed variation in nucleotide sequence corresponds to amino acid variation, nearly all of it being variation among different DNA triplets that code for the same amino acid. That is, any DNA change that causes an amino acid substitution has been weeded out by natural selection, leaving behind only so-called silent DNA variation. In small proteins this weeding-out process has been complete, while in larger enzymes 85 to 95 percent of the amino acid mutations have been removed.

But to be weeded out by natural selection, a change in an amino acid not only must be reflected in the physiology, morphology, or behavior of the organism, but must have an effect on the organism's probability of survival and rate of reproduction. At the moment nothing we know about biochemistry and cell physiology would predict that essentially every amino acid change in, say, alcohol dehydrogenase in *Drosophila pseudo-obscura* should reduce the rate of reproduction of individuals of that species, but the evidence that it happens is incontrovert-

ible. Wherever we look we find much too little protein variation within species, given the amount of DNA variation that
occurs. It seems clear that even the smallest change in the sequence of amino acids of proteins usually has a deleterious effect on the physiology and metabolism of organisms. Thus
the very level where a reductionist methodology is the necessary method of procedure, the determination of the structure
of macromolecules, is the level where we will most surely be
studying entities whose effects usually percolate throughout
the organism and where we must search for the mediation of
those effects.

The qualification "usually" is not to be ignored. Molecular
change in the evolution of a species is often a result of an occasional change in environmental circumstances making a for-
merly deleterious amino acid change favorable. A minority of
changes in proteins have effects so small that random variations in reproductive rate can allow them to increase in frequency in a species and even take over. These "selectively neu-
tral" changes turn out to account for a disproportionately large
fraction of the evolutionary changes *between* species, so that
much of molecular evolution is not the consequence of functional change. Counterintuitively, then, DNA variation between species turns out to be a much less reliable guide to the
organismal level function of molecules than the variation
within species.

≫ In contrast to variations in molecular structure, where small
differences generally have a surprisingly large effect on the
functions of the organism, observed variations in the behavior and morphology of individual organisms in nature usually

have no consistent effect on function when averaged across contexts. It is evident that there is a great deal of variation in shape, size, and activity among individuals in nature. Experiments have consistently shown that much, although not all, of this variation is heritable to some extent, so there is a large amount of standing genetic variation for morphology and behavior within species. It is a basic principle of the action of natural selection that any heritable variation that has an average directional effect on the fitness of organisms will be used up in the process of selection as the genes influencing the trait become fixed in the population. So, if smaller individuals leave more offspring on the average than larger ones, any genes contributing to the size difference will be driven to fixation in the population and there will no longer be heritable variation for size. Any remaining variation in size will be entirely non-genetic in its origin. It follows that the considerable genetic variation for morphology and behavior that is actually observed in nature either is without any effective consequence for the general physiology and metabolism of the organisms or has consequences that vary idiosyncratically from individual to individual or from one environmental context to another in such a way as to make no average difference.

The contrasting conclusions about the likely effect of variations in protein structure as compared to gross anatomy and behavior provide a general guide for drawing preliminary boundaries between quasi-independent subsystems within organisms. The first question in an investigation should be whether there is a great deal of heritable variation for the phenomenon of interest. If there is little or none, then almost surely that phenomenon is effectively functionally linked to a chain of processes that have general consequences for the organism's survival and reproduction. If, in contrast, the phenomenon displays a great deal of heritable variation within the species, then it is likely to be part of some subsystem that is effectively independent of those processes that are vital to the organism. In applying such a heuristic, however, we must take account of the problem of extrapolating from small variations to large ones. The degree to which one property is dynamically linked to others, determining the boundaries of quasi-independent subsystems, may change from one range of values to another. The observed variation in body size within a species may indeed have no average effect on fitness, but an extremely large individual, beyond the bounds of what is ordinarily seen in nature, will have a much reduced fitness. Fruit flies vary in size, but there are no *Drosophila* the size of bumblebees.

≫ The second feature that distinguishes living systems from other physical phenomena is their openness, the characteristic exchange that occurs between the inside and the outside. The stability and reproduction of an organism depend on energetic processes which are possible only if sources of energy outside the organism can be imported into it and if, in turn, the degraded products of energy transformations can be exported to the outside. Even virus particles, which do not metabolize energy, can reproduce only when they become integrated into the metabolic apparatus of the cells that they infect. At the time of viral reproduction there is a complete abolition of the previously existing boundary between the virus and its cellular environment. The softness of the boundary between inside and outside is a universal characteristic of living systems. In Chapter II I made the claim that organisms do not find already existent ecological niches to which they adapt, but are in the constant process of defining and remaking their environments. At every moment natural selection is operating to change the genetic composition of populations in response to the momentary environment, but as that composition changes it forces a concomitant change in the environment itself. Thus organism and environment are both causes and effects in a coevolutionary process.

This coevolutionary process has one general feature that is useful in guiding the direction of its investigation. It is almost always *topologically continuous*. That is, small changes in the environment lead to small changes in the organism which, in turn, lead to small changes in the environment. If this were not true, evolution would have been impossible. If a small change in the organism usually caused a radical and qualitative change in its relation to the outside world, then further continuous evolution of the organism by changes in the frequency of genes could not produce a suitable response. No two successive generations would be selected in the same direction and for the same characteristics. In general the organism and the environment must track each other continuously or life would have long ago become extinct.

The principle of topological continuity is a generality, but it is not a universal law. There are critical junctures in the evolution of organisms, like the phase changes that occur between liquid, gas, and solid states of water as a function of temperature and pressure, at which the relation between organism and environment undergoes a qualitative alteration. Presumably the terrestrial carnivores who were the ancestors of seals spent some time catching their prey at the edges of bodies of water, as bears do now. But a switch occurred in the evolution of those seal ancestors, committing them to a totally aquatic life, while the bears have remained terrestrial. Although in Chapter II I noted many examples of the way in which organisms change their environment, how and when there will be a qualitative change of the environment as a response to changes in the organism is a question to be determined in individual cases. Every case is different, and there are no general principles that can be used to make general predictions about the particular direction of coevolution between organism and milieu.

In order to take proper account of the ordinary topologically continuous changes in the relation of organism and environment, we do not need a revolution in the way experimental observations are made in evolutionary genetics and ecology, only a reorientation of attention. The standard experiments on natural selection are of three kinds. One measures the fecundities and viabilities of different genotypes in a particular environment; the second observes changes in fitness of given genotypes as the environment is altered; the third follows changes in the frequencies of genes in a population over time or differences in the frequencies of genes in different populations as related to the external conditions of the environment. All of these are observations of the response of organisms of different genotypes to particular environmental distributions.

But the same experiments can be carried out in a complementary fashion. We can measure how differences in the genetic constitution of a population change the effective environment of the organisms. This is precisely what was done in the plant engineering experiments described in Chapter II. Even if we cannot measure environmental change directly, we can measure it indirectly through changes in the effect of the environment on the organisms. An example is the series of experiments on *Drosophila*, carried out more than forty years ago, showing that the relative survivorship of larvae of a given genotype is sensitive to the proportions of the different genotypes in the population. When measured in isolation, two genotypes may have equal probabilities of survival, but when each is placed in competition with a third type they will have different relative survivals, the value depending on the proportions of the genotypes in the actual competition. This kind of frequency- and context-dependent fitness may or may not be found in particular cases. We must be prepared for it but not insist on its universality. Again, there is no substitute for actual observation.

In general scientists do what they know how to do and what the time and money available to them allow them to do. New experimental techniques are in part induced by the problems that are under investigation by a community of scientists with common interests, but once those technologies exist they have great power in determining the questions that are asked. The invention of automatic DNA-sequencing machines was a response to a growing demand for sequence, but the availability of such machines and the great ease with which DNA can now be sequenced has meant that the problems on which geneticists work have become those that can be answered from DNA sequences. As there is a dialectic between organisms and their environments, each forming the other, so there is a dialectic of method and problematic in science. Before the mid-1960s, experimental evolutionary genetics included a wide diversity of questions, among them the measurement of fitnesses in nature, the search for genetic variation affecting morphology, physiology, and behavior in natural populations, the evolution of the developmental relation between genetic variation and observed phenotypic variation, and attempts to characterize the amount of genetic diversity within and between populations. The last of these was particularly difficult using the standard techniques of genetic manipulation available to the experimentalist. In order to solve this problem a new technique, protein gel electrophoresis, was introduced, a technique that could be applied to any plant or animal irrespective of whether it could be bred in the laboratory. The result was an almost universal abandonment of the research in all aspects of evolutionary genetics other than the characterization of genetic diversity. A single easily acquired technique changed and pauperized, temporarily it is to be hoped, an entire field of study.

There is nothing in the first three chapters of this book that is not well known to all biologists. Everybody "knows" at some level of consciousness that DNA is not self-reproducing, that the information in DNA sequences is insufficient to specify even a folded protein, not to speak of an entire organism, that the environment of an organism is constructed and constantly altered by the life activities of the organism. But this in-principle knowledge cannot become folded into the structure of biological explanation unless it can be incorporated into the actual work of biologists. Progress in biology depends not on revolutionary new conceptualizations, but on the creation of new methodologies that make questions answerable in practice in a world of finite resources.

NOTES

I. GENE AND ORGANISM

1. A. Rosenblueth and N. Weiner, "Purposeful and non-purposeful behavior," *Philosophy of Science* 18 (1951).
2. D. S. Bendall, ed., *Evolution from Molecules to Men* (Cambridge: Cambridge University Press, 1983).
3. Walter Gilbert, "A vision of the Grail" in Daniel J. Kevles and Leroy Hood, *The Code of Codes: Scientific and Social Issues in the Human Genome Project* (Cambridge, Mass.: Harvard University Press, 1991).
4. Thomas S. Ray Jr., "Growth correlations within the segment in the Araceae," *American Journal of Botany* 73 (1986): 993–1001.
5. J. Clausen, D. D. Keck, and W. W. Heisey, *Experimental Studies on the Nature of Species,* vol. 3: *Environmental Responses of Climatic Races of Achillea.* Carnegie Institution of Washington Publication 581 (1958): 1–129.
6. W. A. Russell, "Comparative performance for maize hybrids representing different eras of maize breeding," in *Proceedings of the 29th Annual Corn and Sorghum Research Conference, Ames, Ia., 1974* (Washington: American Seed Trade Association).
7. A. R. Jensen, "How much can we boost IQ and scholastic achievement?" *Harvard Educational Review* 39 (1969): 15.
8. Gerald Edelman, *Neural Darwinism: The Theory of Neuronal Group Selection* (New York: Basic Books, 1987).

II. ORGANISM AND ENVIRONMENT

1. G. Orians and N. Pearson, "On the theory of central place foraging," in D. J. Horn, R. Mitchell, and G. R. Stairs, eds., *Analysis of Ecological Systems* (Columbus: Ohio State University Press, 1978).
2. Actually, in the region in which I live there are no thrushes of this type, but they exist in other similar grassy places.
3. L. Van Valen, "A new evolutionary law," *Evolutionary Theory* 1 (1973): 1–30.

III. PARTS AND WHOLES, CAUSES AND EFFECTS

1. Although, ironically, individual history is critical for the main function of a clock, since the clock will tell the correct time only if it is set to the correct time when it is started.

IV. DIRECTIONS IN THE STUDY OF BIOLOGY

1. J. Lovelock, *Gaia* (New York: Oxford University Press, 1987), p. 11.
2. See, for example, S. Kauffman, *At Home in the Universe: The Search for Laws of Self-organization and Complexity* (New York: Oxford University Press, 1995).
3. The molecule is changed from an esterase to an organophosphorus hydrolase by the substitution of one amino acid from glycine to aspartic acid. See R. D. Newcombe, P. M. Campbell, D. L. Ollis, E. Cheah, R. J. Russell, and J. G. Oakshott, "A single amino acid substitution converts a carboxylesterase to an organophosphorus hydrolase and confers insecticide resistance on a blowfly," *Proceedings of the National Academy of Sciences U.S.A.* 94 (1997): 7464–68.
4. See G. Gibson and D. S. Hogness, "Effect of polymorphism in the *Drosophila* regulatory gene *Ultrabithorax* on homeotic stability," *Science* 271 (1996): 200–203.

INDEX